F Null

Radio Frequency Radiation

Other McGraw-Hill Books of Interest

Radio Frequency Radiation

Issues & Standards

William F. Hammett, P.E.

McGraw-Hill
New York San Francisco Washington, D.C. Auckland Bogotá
Caracas Lisbon London Madrid Mexico City Milan
Montreal New Delhi San Juan Singapore
Sydney Tokyo Toronto

Library of Congress Cataloging-in-Publication Data

Hammett, William F.
Radio frequency radiation : issues & standards / William F. Hammett.
p. cm.
Includes bibliographical references and index.
ISBN 0-07-025929-1
1. Radio frequency—Safety measures—Standards. 2. Radio frequency—Health aspects. I. Title.
RA569.3.H36 1997
363.18'9—dc21 96-49556
CIP

McGraw-Hill

*A Division of The **McGraw·Hill** Companies*

1 2 3 4 5 6 7 8 9 0 BKP/BKP 9 0 2 1 0 9 8 7

ISBN 0-07-025929-1

The sponsoring editor for this book was Stephen C. Chapman, the editing supervisor was Caroline R. Levine, and the production supervisor was Suzanne W. B. Rapcavage. It was set in Century Schoolbook by Ron Painter of McGraw-Hill's Professional Book Group composition unit.

Printed and bound by Quebecor / Book Press.

This book is printed on recycled, acid-free paper containing a minimum of 50% recycled, de-inked fiber.

This book is dedicated to my children, Toby, Sean, and Reed, in the hope that one becomes a fourth-generation engineer and also helps to harness technology for the benefit of his fellows.

Contents

Preface

This book grew out of an increasing realization, as my consulting engineering firm became more involved as an expert for municipalities and for radiofrequency ("RF") users, that there was not a single, complete source of information about the potential hazards of RF and the regulation of exposure to electromagnetic radiation at RF frequencies. We often found ourselves educating neighbors who were already agitated, as well as municipal planners and telecommunications officers who were already under pressure; in both cases, understanding of certain key facts could have relieved their anxiety and allowed a constructive discussion of the very substantial issues that may, actually, be involved in the specific cases. This book is intended to be that missing source of information.

There is no inherent bias in this book toward any particular side in the many disputes that have occurred, are occurring, and will continue to occur across the United States. These disputes are arising all the more often now that new telecommunications services are being so rapidly introduced and expanded. There is always an underlying scientific truth in such disputes; just as it should make no difference to expert witnesses whether they are retained by municipal governments or a private enterprise, this book strives to present that underlying truth, as best as I can recognize it at the present time.

The reader will find material in this book that should satisfy the need for information at several levels: basic technical information about the nature of RF and its interaction with the human "receiver," further information about the prevailing standards and their development, and supplementary information about the assessment of RF exposure in particular cases and the possible mitigation measures that might be taken. Since this is a growing field, consideration is also made of current trends in the areas of exposure research, regulation of RF sources, and development of new devices. What is learned here can continue to be applied to specific situations in the future.

The discussion begins at the "lay" level. Much as classical, Newtonian physics is useful for describing virtually all phenomena that are directly observable, I have tried to use descriptions that make sense in virtually all settings, but with which a physicist could quibble. My goal was to avoid prose descriptions that are as dense as the subject matter, and I trust that other experts will indulge this approach.

Special note to engineering professionals. Due to the need for this book among a broad audience, Chapter 1 includes some very basic science, and you will probably want to skip most of it. The science underlying this field of inquiry has been learned during decades of schooling, graduate work, and private practice; further discussion of the basics can be found elsewhere, of course, but they are summarized here to allow this text to stand on its own. Nevertheless, I might direct your attention to part of the Exposure and Dosage section in Chapter 1, and all of the section on Biological Effects in that chapter, if you are not already familiar with this aspect of the field.

Organization and Conventions

This text is organized into five chapters, each consisting of several sections that deal with a distinct aspect of the radiation phenomena or the regulatory environment. This organization is designed to enable the lay reader to find the material of immediate interest without necessarily knowing exactly what entry to look up in the index, and it will enable that reader to understand the context of the material by reading the whole section (or chapter).

Boldface type is used in this book to identify new terms, while *italics* is used for emphasis, and underlining for clarity. In light of the fact that different permitting agencies, whether the Federal Communications Commission or the local town elders, may adopt different standards for assessment of RF exposure conditions, reference in this book is often made to the "PS" (*p*revailing *s*tandard) or the "Standard," for which one can substitute any of the several standards in use. In those situations dealing with differences among the standards, of course, they will be individually identified.

Many of the current exposure standards utilize a two-tier approach, in which different guidelines are adopted either for public and occupational exposure situations or for uncontrolled and controlled environments. The distinctions between "public" and "uncontrolled" and between "occupational" and "controlled" are so subtle as to be nonexistent. Therefore, the terms "public" and "uncontrolled" are used interchangeably, as are "occupational" and "controlled."

For ease of comparison among studies, standards, and applications, only a single unit of power flux density is generally used: mW/cm^2. Most foreign organizations favor W/m^2, and sometimes μW/cm^2 is more convenient, but the intent here is to enable understanding of the issues and of the relationships between studies and standards, and between standards and applications. A single measure of this important parameter should help.

Footnotes are used to provide more detailed explanation or to share interesting information that may not be essential to the main line of reasoning in the body of the text. Because of the general nature of so much of the material, references are simply listed at the end of the book.

The application of this book is to matters relating to radio frequencies (RF), and my own expertise is limited to that frequency range. It may be appropriate to comment on related or bounding fields of inquiry, in order to identify or clarify the RF matters under discussion, but no claim of expertise is made in those other fields. Specifically, that includes the study of extremely low frequency electromagnetic fields, as well as the study of ionizing electromagnetic fields. These are considered separate fields of inquiry, and their effects on human beings are acknowledged to be different. In addition, since matters relating to the topic of this text can become subject to litigation, it must be disclaimed that the author is not an attorney, and no part of this text constitutes a *legal* opinion offered by the author, even in direct application to compliance with regulations having force of law.

Special note regarding historical perspective. A technical manual in the methods for complying with some standard need not be concerned with the development of that standard. This book, however, is more than a technical manual, and its readers may want to understand the "why" as well as the "how." Therefore, in addition to the underlying science found in Chapter 1, the history of RF regulation is reviewed in Chapter 2. The debate about RF exposures needs to be set in context; in the reader's vocational capacity, especially, it may be beneficial to have a working familiarity with the early standards and their revision, with the standards in use in other countries, and with the several parties in the United States affecting the promulgation and adoption of new standards.

The question of evaluating the safety of technological developments becomes easily politicized. That has certainly been the case with deployment of new RF systems. As the consumer marketplace demands greater amounts of RF spectrum for communication services, and as the FCC makes more spectrum available, the likelihood increases of battles for public opinion. In fact, those battles are being waged now, from which has developed the need for this book.

Nor is the regulatory situation static. As battles are fought at local levels, new regulation results. The U.S. Congress has weighed in heavily with the preemption language of the Telecommunications Bill of 1996, discussed in Chapter 5, but local battles continue with only slightly changed focus. RF exposure issues remain controversial.

It is hoped that the information and analyses found in this book will encourage both the prudent deployment of RF emitters and their wise regulation.

Acknowledgments

I gladly acknowledge my father, Robert L. Hammett, whose quiet encouragement of my technical inclinations helped lead me to study enginering, whose willing acceptance of me into the firm he founded gave me the opportunity to specialize in this field, and whose wise guidance has enabled me to lead that firm after his retirement. I also wish to acknowledge the strong influence of my mother, Luana, from whom I acquired an appreciation for the written word. I am especially grateful to my wonderful wife, Sara, who patiently awaited my return from the long period of solitude needed for writing.

The entire staff at Hammett & Edison, Inc., Consulting Engineers, lent support in many ways, especially with the location of obscure references, the production of the many illustrative figures, the critiquing of early drafts, and the proofing of the manuscript.

Final thanks go to k.d. lang, whose continual serenade during the many hours of writing kept me focused on the task at hand.

William F. Hammett

Chapter

1

Underlying Science

EMF Spectrum

Electromagnetic fields are a natural part of all life, providing light, heat, and, relatively recently, power and communications. Also known by the acronym **EMF,** such fields exist in nature in many forms, and they are also created, often intentionally but sometimes as a by-product, by much of modern technology.

Correlated physical manifestations of a particle wave, electric and magnetic fields are created by the flow of electric or magnetic energy. For instance, the long spinning of the earth has created an alignment of iron-bearing molecules and the presence of magnetic north and south poles. Magnetic energy flows from one pole to the other, and movement of a conductive wire through that magnetic field will cause electricity to flow in the wire.

A common battery, such as those sized from AAAA to D, stores energy. All batteries have two ends or connection points, which are often shaped differently to help distinguish one from the other. One end is known as positive and the other as negative. Connecting the two ends with a wire lets energy (in the form of electricity) flow from one end to the other. Putting some electrical device between the two ends, instead of just a connecting wire, applies the energy to that device and thus some useful work is performed. Whether the device is a camera, flashlight, toy car, or any one of thousands of other battery-operated goods, it derives its energy from the flow of electricity between the ends of the battery.

The flow of electric current along a wire creates a corresponding magnetic field surrounding the wire, as shown in Fig. 1.1. In certain cases, such as with an electromagnet, this corresponding field is the sought-after goal, but in most cases it is simply a side effect of the

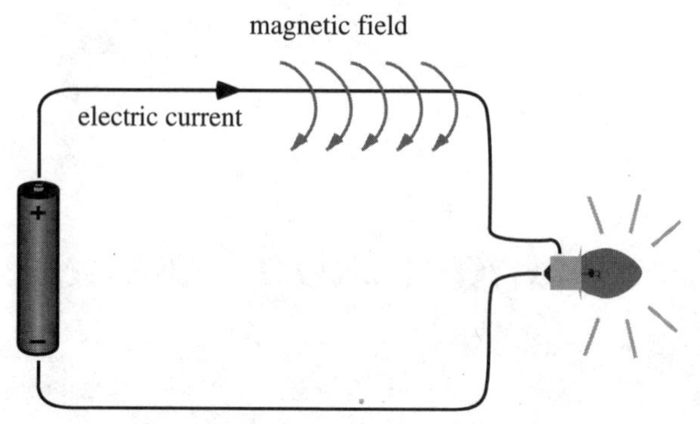

Figure 1.1 Electric current and magnetic field.

principal goal, the flow of energy to some other location or to some electric device.

The ends of a battery are known as **poles,** and one has to connect a battery with the proper **polarity** to provide the current flow needed by the electrical device so that it may be powered. Such flow is called **direct current,** flowing constantly as it does from positive to negative.[1] However, if the polarity changes (*i.e.,* if the positive end became negative and the negative end positive), the electric current flows in the other direction. If such a change in polarity repeats regularly, the electric flow is said to have a **frequency** equal to the number of times that the polarity changes during some unit of time. When the speed with which the polarity changes is increased, the characteristic frequency is increased. Direct current does not change polarity and therefore has no frequency.

The field emanating from a current that has changing polarity also has a changing polarity. Figure 1.2 shows the amplitude of a fixed-polarity field, represented by a line of constant amplitude. Also shown is the (strength) of a changing-polarity field; it equals the fixed-polarity amplitude at certain instants, decreasing to a negative amplitude of the same value and then returning to maximum positive amplitude to form a complete cycle. This is the **wave** phenomenon characteristic of changing-polarity fields and follows the pattern of the trigonometric sine wave.

Recall that the flow of electricity creates a corresponding magnetic field; changing-polarity fields create a composite wave of both electric and magnetic energy, as shown in Fig. 1.3. The composite field created by this wave is called an **electromagnetic field,** or **EMF.** A useful characteristic of EMF is that a conductor placed in such a field will

---

[1]Some people say that the "actual" current flows from negative to positive.

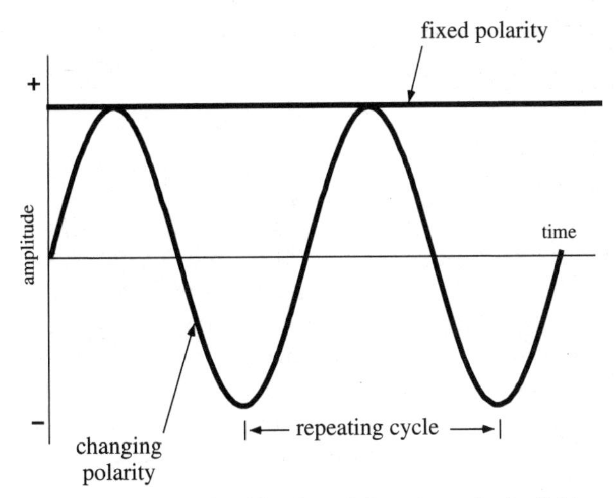

Figure 1.2 Amplitude of fixed- and changing-polarity fields.

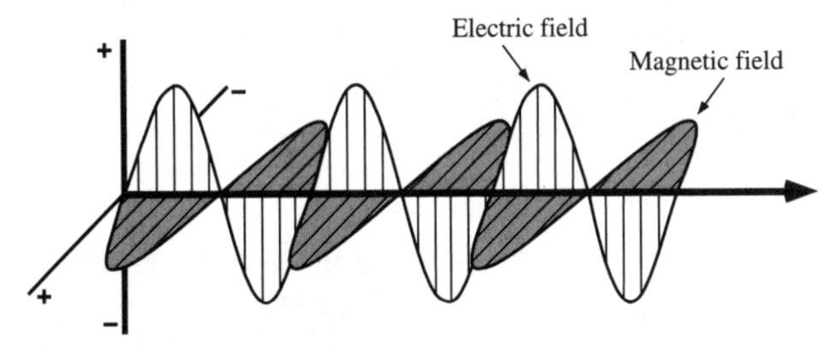

Figure 1.3 Composite electromagnetic wave.

have an electric current **induced** in that conductor, converting some of the energy from the EMF back to energy flowing in the conductor. The frequency of this induced current will match that of the current in the conductor that created the EMF. This is the basis for using EMF as a means of wireless communication and is the primary reason for the deliberate creation of EMF. A radio or TV receiving antenna is a specialized conductor designed to receive a signal via the induced energy from an EMF radiated from a transmitter. (Of course, *any* conductive object placed in an EMF will have some current induced, even when that is not the desired result. This book is concerned with the situation that arises when that object is a human body, imperfect conductor though it may be.)

Electromagnetic fields can exist with characteristic frequencies over an enormous range. The chart in Fig. 1.4 shows this EMF **spectrum** of frequency, which can be characterized in a variety of different ways. This particular chart shows a grouping of EMF by common human references. Electrical power is distributed over wires as cur-

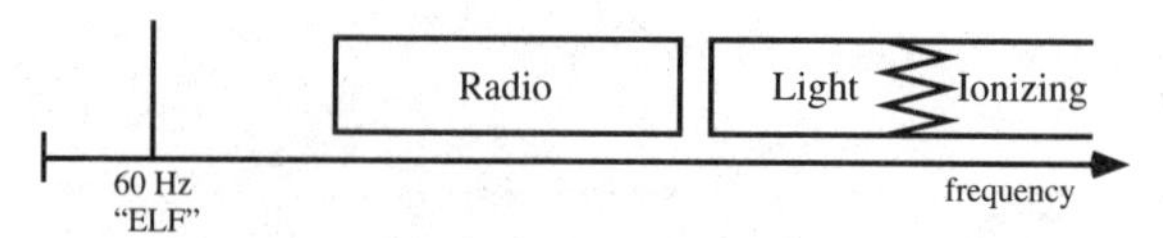

Figure 1.4 Divisions of EMF spectrum.

rent that alternates and restores its polarity 60 times a second; this is considered an **extremely low frequency,** or **ELF** (*not to be confused with the similar-looking term EMF*). The three ranges of EMF shown are radio, light, and ionizing. **Radio** is a general term applying to the use of EMF for wireless communications of many types, not just AM or FM radio. Included among a myriad of wireless communications services is amateur ("ham"), citizens band, television, cellular telephone, and microwave. **Light** is that part of the EMF spectrum to which human eyes respond, plus a little on each side. In fact, visible light covers a remarkably narrow range of frequencies. Light below that visible to humans includes infrared, felt by humans as radiated heat, and light above the visible range includes ultraviolet, which can cause sunburn. Frequencies above light are known as **ionizing** frequencies, in recognition of their ability to enter atoms and break electron bonds.

The term **radiation** applies to the dispersal of EMF energy. Once generated, EMF radiates in all directions, depending on how it may have been focused, intentionally or not. As it disperses, the effective power of the EMF is dissipating, and the EMF is reflected or absorbed to varying degrees as it comes into contact with different types of matter.

Units and Definitions

Essential to a common understanding in any endeavor is a common set of definitions. The scientific community has traditionally used two different systems of units when describing physical phenomena: the **English system** and the **metric system.** The English system uses feet for distance (along with inches, yards, and miles) and pounds for mass. The metric system features unit descriptions that conveniently vary by multiples of 10 and is the increasingly common standard, even among the English-speaking nations. The metric units can be either *mks,* using meters (length), kilograms (mass), and seconds (time) as the basic units for other definitions (*e.g.,* 1 newton of force is that required to accelerate 1 kilogram of mass at 1 meter per second per second), or *cgs,* using centimeters, grams, and seconds (*e.g.,* 1 dyne of force accelerates 1 gram at 1 centimeter per second per second). For the purposes of this text, the metric mks system, also called the **Système International d'Unités (SI),** will be used most often, but conversions to other unit systems are easily developed. The use of certain English units will also be made occasionally, since they are still commonly understood by lay readers and lay audiences.

Distance, mass, and time are intuitive concepts, described in units of meters, kilograms, and seconds. Other concepts that need to be introduced and defined are energy, power, power density, radiation, specific absorption rate, and order of magnitude.

Energy is the ability to do work, and it can exist in many different forms. Energy can be stored, as electrical energy is stored (by chemical means) in batteries. Other examples include storing potential energy by compressing a spring or by hoisting a weight on a pulley; these represent conditions where work can be performed by the release of that energy. The SI unit of energy is the **joule (J),**[2] defined as the energy stored (or expended) by the force of 1 newton acting over a distance of 1 meter (1 m).

Power is the rate at which energy is consumed (or produced) (*i.e.,* power = energy ÷ time). The SI unit for power is the **watt,**[3] defined as 1 joule per second. A 100-watt (100 W) light bulb, for instance, uses 100 W of electrical power, and a 60-horsepower outboard motor can produce 44,700 W of mechanical power.

Power density, also called **power flux density,** is a *distribution* of power over some area. There is no unit specially defined for power density, as with newtons, joules, and watts for force, energy, and power. Thus, power density is expressed simply in units of power per area, such as watts per square meter (W/m^2), milliwatts per square centimeter (mW/cm^2), or microwatts per square centimeter ($\mu W/cm^2$).

Radiation is the dispersal of energy from a source.[4] For instance, when a 100 W light bulb radiates EMF in the form of light and heat, it is dispersing the 100 W of electrical power that it is consuming. The dispersal occurs roughly equally in all directions, and the power density for the operation of the light bulb can be easily calculated. The power radiated at a single instant leaves the bulb and travels outward at a constant velocity, forming a sphere that expands as the energy travels away from the bulb. When the sphere has grown to 1 m in radius, its surface area is $A = 1.33\ \pi\ r^2 = 4.2$ square meters (m^2). Since

---

[2]Named after James Prescott Joule (1818–1889), an English physicist.

[3]Named after James Watt (1736–1819), a Scottish engineer and inventor.

[4]The term **radiation** should not be frightening. It does *not* imply radioactivity, which is the radiation of atomic particles due to the spontaneous decay of an unstable substance. Light and heat from an incandescent bulb are radiation, the ripples in a pond from a dropped pebble are radiation, and the sounds of speech are radiation, these last two being examples of radiation that is not electromagnetic. Radiation simply means dispersal.

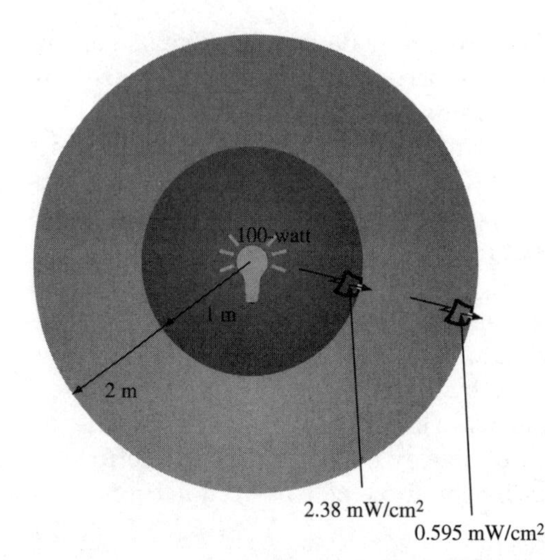

Figure 1.5 Power flux density.

the 100 W is distributed evenly over that sphere, the power density is 100 W ÷ 4.2 m^2 = 23.8 W/m^2 (2.38 mW/cm^2), as shown in Fig. 1.5.

When that sphere of radiation has grown to 2 m, the 100 W will be distributed over an area $A = 1.33\ \pi\ r^2 = 16.8$ m^2, for a power density of 100 ÷ 16.8 = 5.95 W/m^2 (0.595 mW/cm^2). Thus, the density of the radiated power depends on the distance from the radiation source; intuitively, this should be clear. It is also important to note that when the distance was doubled in the example, from 1 m to 2 m, the power density went down by a factor of 4 (23.8 ÷ 4 = 5.95), not just 2. This is because the surface of the expanding sphere increases by the *square* of its radius. This understanding is so fundamental to physical science that it is called the **Inverse Square Law:**

> *The power received is inversely proportional to the square of the distance from its source.*

For an increase in distance of 10 times, therefore, the power density must drop by a factor of 100; at a distance 100 times greater, the power density drops by 10,000 times.

Specific absorption rate is, as the term suggests, the rate at which energy is absorbed by a unit of biological tissue. Also known by its acronym **SAR,** its units are therefore energy ÷ time ÷ mass, or joules per second per kilogram. Since a watt is a joule per second, SAR is normally expressed in units of watts per kilogram (W/kg).

The term **order of magnitude** refers to an estimation of size relative to a reference. It is an exponential relationship, whereby a hundredfold increase would be considered a two-order of magnitude increase ($10^2 = 100$), and a thousandfold increase (or decrease) would

prefix	*abbr.*	*multiplier*	*meaning*
tera	T	10^{12}	trillions
giga	G	10^{9}	billions
mega	M	10^{6}	millions
kilo	k	10^{3}	thousands
		$10^{0} = 1$	**basic unit**
centi	*c*	10^{-2}	*hundredths*
milli	m	10^{-3}	thousandths
micro	μ	10^{-6}	millionths
nano	n	10^{-9}	billionths
pico	p	10^{-12}	trillionths

Figure 1.6 Metric prefixes.

be three orders of magnitude greater (or lesser). Thus, anything referred to as an order of magnitude change represents a difference of at least 10 times, and a change of several orders of magnitude represents a really big change.

One advantage of the metric system is the ease with which the units can be adjusted to fit quantities ranging for many orders of magnitude. For instance, with a basic unit of a meter, we frequently encounter *kilo*meters (thousands of meters) and *centi*meters (hundredths of a meter), which are convenient units five orders of magnitude apart. Figure 1.6 shows the standard metric multiplier prefixes, spaced at three orders of magnitude. (Centi- does not fit that spacing but is included because it is also common.) These prefixes can be applied to any fundamental unit, to give such useful units as milliliters (ml, thousandths of a liter of volume), megawatts (MW, millions of watts of power), picofarads (pf, trillionths of a farad[5] of capacitance), and nanometers (nm, billionths of a meter of length), as well as such new units as gigabytes (GB, billions[6] of bytes of computer storage).

The final unit to define is a **decibel (dB).** This term was developed from a measure of sound, the bel,[7] and reflects the fact that the human ear has a logarithmic response. A decibel is a unitless ratio of two powers, defined by the expression:

$$\text{dB} = 10 \log_{10}\left(\frac{P_1}{P_2}\right)$$

Thus, the following relationships between P_1 and P_2 can be expressed in decibels as follows:

[5]Named after Michael Faraday (1791–1867), an English physicist and chemist.

[6]Actually, kilo in computer jargon is 1,024 (2^{10}), and a gigabyte is 1,073,741,824 bytes (2^{30}).

[7]Named after Alexander Graham Bell (1847–1922), a U.S. scientist and inventor.

$$P_1 \text{ is 1,000 times greater than } P_2 \rightarrow 30 \text{ dB}$$

$$P_1 \text{ is 100 times greater than } P_2 \rightarrow 20 \text{ dB}$$

$$P_1 \text{ is 10 times greater than } P_2 \rightarrow 10 \text{ dB}$$

$$P_1 \text{ equals } P_2 \rightarrow 0 \text{ dB}$$

$$P_1 \text{ is 10 times smaller than } P_2 \rightarrow -10 \text{ dB}$$

$$P_1 \text{ is 100 times smaller than } P_2 \rightarrow -20 \text{ dB}$$

$$P_1 \text{ is 1,000 times smaller than } P_2 \rightarrow -30 \text{ dB}$$

Frequency and Wavelength

Frequency

Frequency refers to the regular recurrence of an event, such as moment of maximum magnitude of an EMF wave. Typical units of EMF frequency are **hertz,**[8] abbreviated Hz, which is the number of repeating **cycles per second.** As another example, consider July Fourth, which is an annual holiday; its frequency is once per year. Similarly, Friday has a frequency of once per week, and a cuckoo clock might chime once per hour. Note that the size of the event is *not* described by its frequency: July 4 means nothing in Paraguay, but its frequency is still once per year, and the cuckoo clock might be loud or soft, but its chiming frequency is still once per hour. Likewise, the frequency of EMF provides no indication of its magnitude. A 54 MHz TV station could operate with 100 kW radiated power, while a 2,500 MHz TV station (50 times higher in frequency) might operate with only 100 W radiated power (1,000 times less power).

Frequency is a key distinguishing feature of EMF. The EMF spectrum graph in Fig. 1.7 is based on frequency, with the low-frequency part of the spectrum beginning at a single cycle per second on the left and continuing on to frequencies so high that we are only now learning to detect them. (Sound, although it is shown on the spectrum chart, is not really part of the electromagnetic spectrum. Sound energy is carried by pressure waves that form in air or some other medium. Electromagnetic energy is carried by photon waves that do not require a medium for transport. Thus, sound is an earth-bound phenomenon, while EMF literally radiates throughout the universe. Nevertheless,

---

[8]Named after Heinrich Rudolph Hertz (1857–1894), a German physicist.

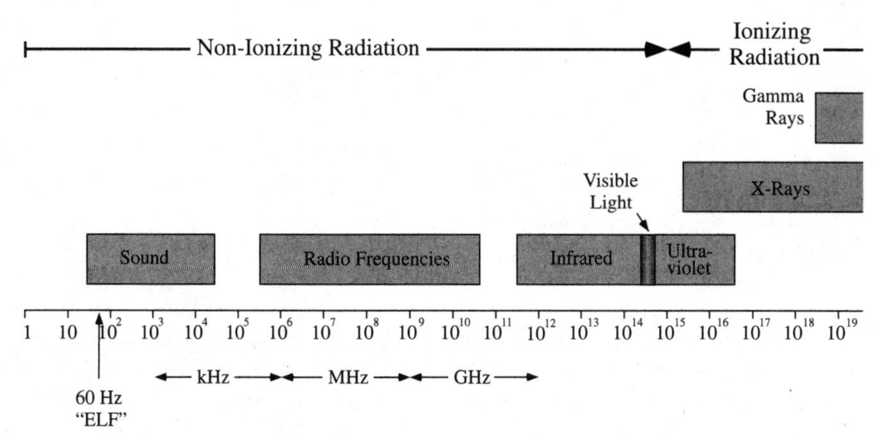

Figure 1.7 EMF spectrum.

representative frequency	*band designation*	*abbr.*	*frequency range*
100 Hz	Extremely Low Frequency	ELF	30–300 Hz
1,000 Hz	Voice Frequency	VF	300–3,000 Hz
10 kHz	Very Low Frequency	VLF	3–30 kHz
100 kHz	Low Frequency	LF	30–300 kHz
1,000 kHz	Medium Frequency	MF	300–3,000 kHz
10 MHz	High Frequency	HF	3–30 MHz
100 MHz	Very High Frequency	VHF	30–300 MHz
1,000 MHz	Ultra High Frequency	UHF	300–3,000 MHz
10 GHz	Super High Frequency	SHF	3–30 GHz
100 GHz	Extremely High Frequency	EHF	30–300 GHz

Figure 1.8 EMF band designations.

the frequencies of pressure waves audible to humans are shown for interest.)

Figure 1.8 lists the various frequency band designations that have developed over the years, with a new designation being assigned to each tenfold increase in frequency. Some of the abbreviations may seem familiar, such as VHF and UHF. These two terms have been applied to television channels that happen to fall within those bands.

The use of **radio frequencies (RF)**, customarily considered to include frequencies from 300 kHz up to 300 GHz, is shown in more detail in Fig. 1.9. These frequencies have been actively exploited since the 1930s for wireless communications purposes, and the relative scarcity now of such frequencies has caused reassignments to occur. For instance, the old UHF television Channels 70 through 83 were taken out of service in 1974 and reassigned, in part, for use by the emerging cellu-

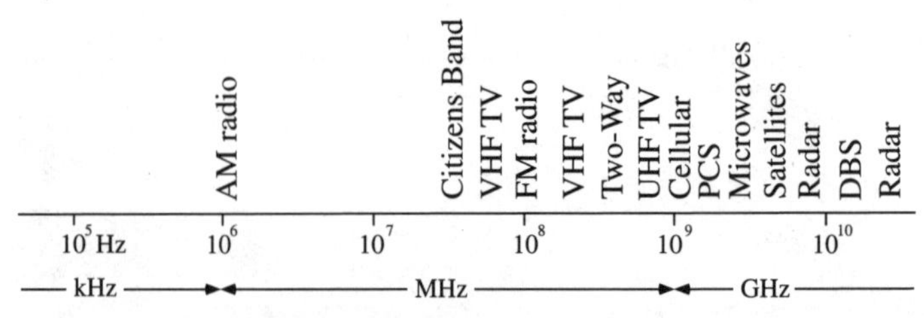

Figure 1.9 Selected radio frequencies.

lar telephone industry. The **Federal Communications Commission**[9] (**FCC**) retains jurisdiction over civilian use of all RF frequencies; frequency sharing and realignment is a common aspect today of the FCC's regulatory activities.

RF versus 60 Hz

A distinction needs to be drawn between radio frequencies (RF) and extremely low frequencies (ELF). The most common ELF frequency is 60 Hz, this being the power-line frequency used in the United States.[10] One sometimes hears the comment that, "It's *all* EMF—if it's dangerous at one frequency, it's dangerous at them all." Saying that EMF is all "the same" is like saying that all chemical elements are the same since they all have protons, neutrons, and electrons. While at some level of understanding they probably *are* the same, there are, in fact, obvious and meaningful differences between, for instance, helium and iron. Likewise, there are critical distinctions that must be made among various portions of the EMF spectrum.

It has become common practice for electric utility companies to include in occasional billing cycles a flyer discussing EMF and recommending "prudent avoidance." Since research findings are more definitive as to potential health effects at RF, rather than at 60 Hz, it is often important at public meetings regarding the siting of new RF transmitting facilities to draw analogies to common numerical references in order to help lay persons grasp how disparate RF is from 60 Hz. Two favorites, one of which it is hoped an audience will respond to, involve money and speed.

Imagine, for a moment, having $60 in your pocket; you know what that feels like. Then imagine having $6 (10 times less), which will buy only a quick meal, or having $600 in your pocket, which could buy

[9]Created by Congress in 1934 and operating as an independent agency.

[10]The power-line frequency in Great Britain, Europe, and many other regions is 50 Hz.

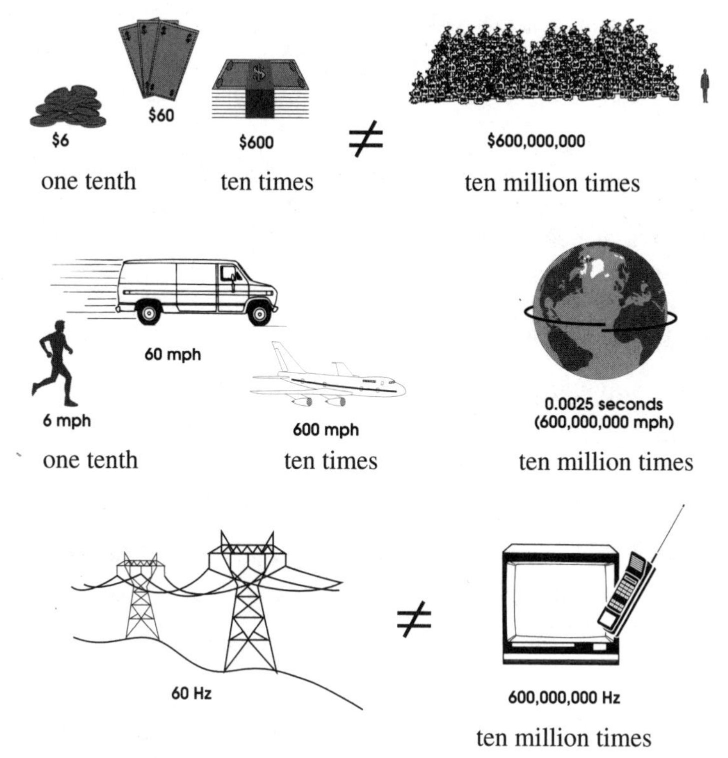

Figure 1.10 There's a difference!

much more, even an old car, perhaps. This generally feels like a big difference, that between $6 and $600, 10 times below and 10 times above the $60 you first considered. Now think of $6,000 (100 times above $60), then $60,000 (1,000 times), and $600,000, and $6,000,000, and $60,000,000, and finally $600,000,000. That is six hundred million dollars, the virtual equivalent of winning the lottery in every state. That's how big the difference is between 60 Hz, the power-line frequency, and 600 MHz, a typical RF frequency. Figure 1.10 illustrates this comparison, as well as another helpful one.

Or imagine that you are in a car traveling on a freeway at 60 miles per hour (mph); you know what that feels like. Now imagine traveling only 6 mph, 10 times less; that's the speed of a comfortable jog. Or imagine traveling 600 mph, 10 times more; that's the speed of a commercial jetliner. That's a big difference, between jogging across the country, say, and flying there in 5 hours. Now think of traveling 600 million mph—you'd circle the earth in less than a hundredth of a second (literally, the blink of an eye); in fact, you could get to the sun in 9 minutes (9 min), almost as fast as light itself, which takes 8 min for that trip. This is another example of how big the difference is between the 60 Hz power-line frequency and RF frequencies.

Wavelength

Frequency was defined by the *time* between recurring regular events. All EMF travels in space, not surprisingly, at the speed of light.[11] Thus, **wavelength,** which is the *distance* that the wave travels during that time, can be determined for EMF just by knowing the frequency. This relationship between wavelength, often denoted by the Greek letter lambda (λ), and frequency is given by

$$\text{wavelength} = \text{speed} \div \text{frequency},$$

which is equivalent to:

$$\text{frequency} = \text{speed} \div \text{wavelength}.$$

For frequency in megaherz and wavelength in meters, the formulas become

$$\lambda = \frac{300}{f} \quad \text{and} \quad f = \frac{300}{\lambda}$$

Wavelength is the second distinguishing characteristic of EMF. Clearly, frequency and wavelength have a reciprocal relationship (*i.e.,* when one goes up, the other goes down). Figure 1.11 illustrates this, with a lower frequency having a longer wavelength and a higher frequency having a shorter wavelength.

The wavelength of visible light ranges from 700 nanometers (red) to 400 nm (violet), with the rainbow of colors at intermediate wavelengths. The wavelength for RF ranges from 1 kilometer (1 km) at 300 kHz to 1 millimeter (1 mm) at 300 GHz; perhaps foretelling some of the findings reviewed later, the wavelength at VHF varies from 10 to 1 m, a range that includes the length of the human body. The wavelength for 60 Hz, on the other hand, is 5 million meters! This length is 5,000 km, which is about the breadth of the continental United States and is intuitively different from wavelengths at RF broadcast frequencies.

Energy

The energy that an EMF wave carries is the third distinguishing characteristic of EMF. The behavior of electromagnetic waves at ion-

[11]Approximately 300,000,000 meters per second.

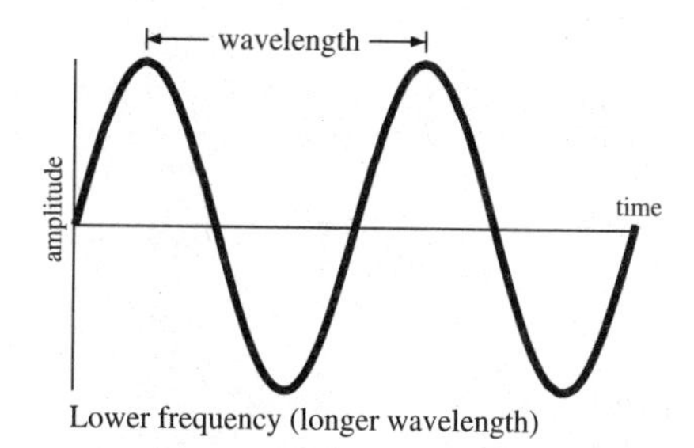

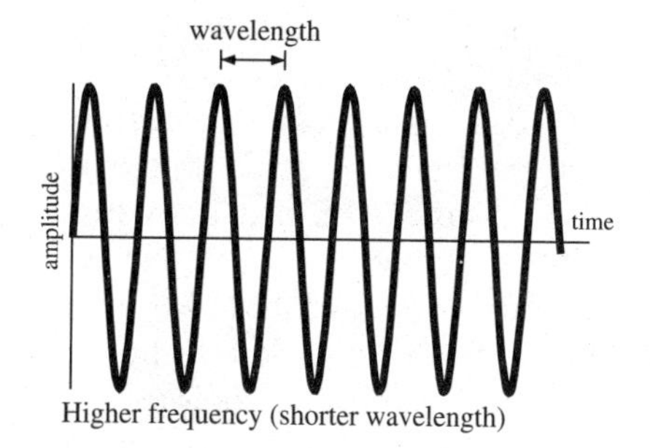

Figure 1.11 Frequency and wavelength.

izing frequencies, where the wavelength is on the order of atomic dimensions, is like a stream of particles (called **photons**). Each EMF has a characteristic **quantum energy** (also called **photon energy**), given by the equation

$$\begin{aligned} QE &= hf \\ &= 4.135 \times 10^{-15} f \text{ electron-volts,} \end{aligned}$$

where h is Planck's constant[12] (6.624×10^{-27} erg-seconds) and f is the frequency of the wave, in hertz (cycles per second).

Research in chemistry and physics has indicated that quantum energies in excess of about 4 electron-volts are required to break an electron free from an atom of matter, a process known as **ionization.** Therefore, EMF at frequencies above about 10^{15} Hz is capable of this, and so is known as **ionizing radiation.** This portion of the spectrum includes some ultraviolet light, X rays, gamma rays, and cosmic rays. ***Non*ionizing electromagnetic radiation (NIER)** is characterized, by definition, by quantum energies *below* the ionization threshold and therefore cannot break atomic bonds *regardless of the power density.* Thus, exposure to NIER, below about 10^{15} Hz, is not necessarily de-

---

[12]Named after Max Karl Ernst Planck (1858–1947), a German physicist.

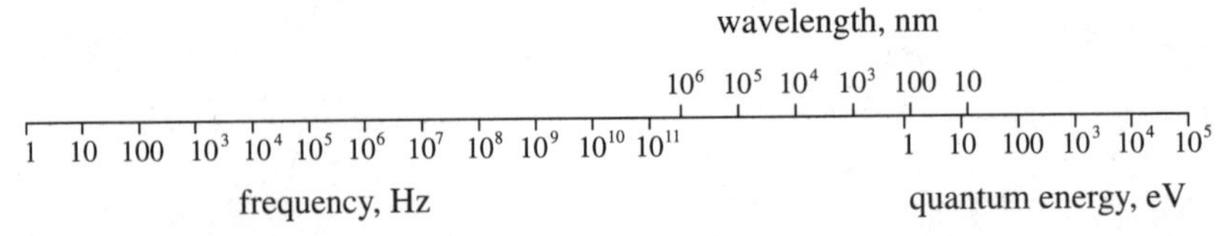

Figure 1.12 EMF spectrum by appropriate units.

structive or cumulative. This distinction between ionizing and nonionizing EMF is extremely important, and it is universally recognized among scientists and the standards-setting bodies who review their work.[13]

Another way to consider EMF energy, less rigorously but perhaps more intuitively, is by its wavelength. Atoms are tiny structures, with diameters on the order of one angstrom (1 Å) unit.[14] Since 1 Å equals 10^{-10} m (0.1 nanometers), frequencies above about 3×10^{18} Hz will have wavelengths small enough to get inside an atom and break loose an electron. This is ionizing energy. At frequencies below light, the EMF has wavelengths of at least 8000 Å (*i.e.,* too big to fit into individual atoms); this is nonionizing energy. In fact, at 2 GHz, a common microwave frequency, the wavelength is 1.5×10^9 Å, or about 6 inches (6 in).

There are, thus, three ways to characterize EMF: by frequency, by wavelength, and by quantum energy. They are all related, of course, but they serve as convenient ways of dealing with disparate sections of the EMF spectrum. This may represent the most useful way of considering the *entire* EMF spectrum, as shown in Fig. 1.12. The three segments, known by their frequency, wavelength, or quantum energy, roughly correspond to the segments identified as radio, light, and ionizing in Fig. 1.4.

Exposure and Dosage

A separate characteristic of EMF is its **amplitude.** As noted earlier, the strength of RF radiation is independent of its frequency. A radio signal at a given frequency may have low or high **field strength** at some location; a higher-frequency signal may have a lower field strength, and vice versa.

In a scientific context, **exposure** is the instantaneous presentation of some factor. Note that exposure does not imply any particular

[13]The American Conference of Governmental Industrial Hygienists treats 12.4 electron-volts as the ionization threshold, corresponding to about 3×10^{15} Hz.

[14]Named after Anders Jonas Ångström (1814–1874), a Swedish astronomer and physicist.

amount of absorption of that factor, merely its presence. **Dosage** is the actual quantity *received* of that factor; it is dependent on 1) the **susceptibility** to absorption of the matter exposed and 2) the **duration** of the exposure. For instance, placing a black object in full sunlight for an instant will not result in significant temperature rise, since the duration is so short. Similarly, placing a mirrored object in sunlight, even for an extended period, will not cause its temperature to rise significantly since its susceptibility to absorption is so low. Of course, placing a black object in sunlight for an extended period fulfills both the susceptibility and duration requirements, and its temperature will rise.

This relationship is described by the formula

$$\text{dosage} = \text{susceptibility} \times \text{exposure magnitude} \times \text{duration.}$$

The exposure magnitude for RF radiation can be expressed in power density units, such as W/m^2. The susceptibility is related to the body area subjected to the RF radiation, which might be expressed in m^2. The equation then becomes

$$\text{dosage} = \text{body area} \times \text{power density} \times \text{duration}$$

$$J = m^2 \times W/m^2 \times \text{seconds}$$

For a specific unit mass, this becomes

$$J/kg = W/kg \times \text{seconds}$$

which is recognized as

$$SA = SAR \times \text{duration}$$

$$\text{specific absorption} = \text{specific absorption rate} \times \text{duration}$$

The susceptibility for any particular material is dependent on its **conductivity** (ability to conduct energy) and **permittivity** (ability to store energy). Determination of these parameters has been made for many materials, but, for the purposes of this text, that work need not be reviewed. The concern here is with the impact on biological tissue, the conductivity and permittivity of which falls within a limited range compared with other materials. To the extent that these parameters are frequency dependent, which they are expected to be, that dependence will be reflected in the experimental data taken, for which frequency is treated as an independent variable.

A major parameter affecting the susceptibility for absorption is the *size* of the irradiated body relative to the wavelength of the irradiating field. As noted in Figure 1.13, when the wavelength is much larg-

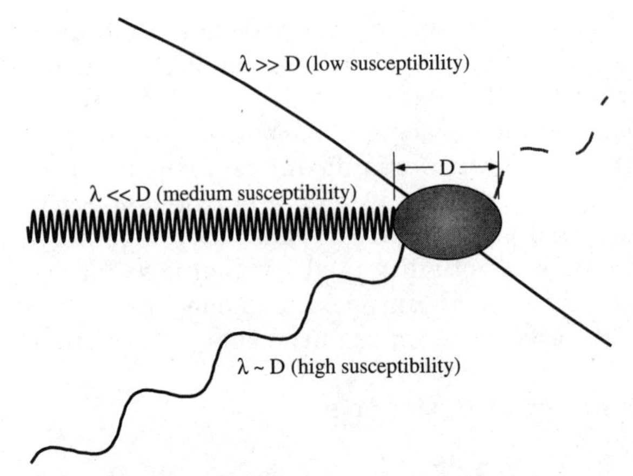

Figure 1.13 Susceptibility versus wavelength.

er than the body, susceptibility is low. When the wavelength is much shorter than the long dimension of the body, susceptibility is somewhat higher and is dependent on the depth of penetration of the RF energy. Finally, when the wavelength is on the order of the long body dimension, a **resonant condition** is approached, and the susceptibility to absorption of RF energy is highest. Extreme examples of these conditions would be:

- 60 Hz (low susceptibility), where the wavelength of 5 million meters (*i.e.*, New York to San Francisco) is many orders of magnitude greater than the length of the human body
- Light (medium susceptibility), where the wavelength of about 0.5 micrometers (0.5 μm) is many orders of magnitude less than a human
- FM radio (highest susceptibility), where the wavelength of 3 meters is about twice the typical human height

For bodies irradiated at long wavelengths, absorption is poor and thus SAR is not a good indicator of whatever interaction occurs. Stimulation via induced electrical currents is the primary mechanism of interaction.

For irradiated bodies near resonance, a major parameter affecting the susceptibility for absorption is the **alignment** of the body with respect to the wave propagation front. As shown in Fig. 1.14, an irradiated body can be aligned with the electric field (E alignment), with the magnetic field (H alignment), or with direction of propagation (k alignment). Experiments have shown that the greatest coupling to an elongated body occurs when the long axis of the body is aligned with the

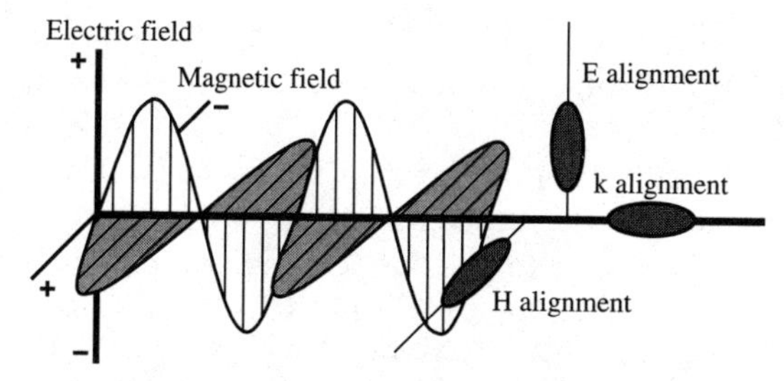

Figure 1.14 Possible alignments of irradiated body.

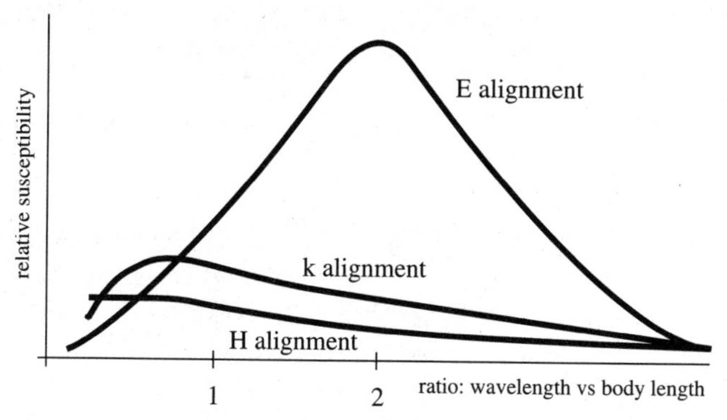

Figure 1.15 Effect of alignment with EMF.

electric field and, further, that the body is most absorptive when the wavelength is about twice the long axis length.[15] The nature of this relationship is reflected by the chart in Fig. 1.15, which shows the slightly greater absorption of energy in the k alignment than in the H alignment but the markedly greater coupling in the E alignment.[16]

For bodies irradiated at short wavelengths, absorption is limited to surface areas, and SAR is again not a good indicator of whatever interaction might result.[17] Surface heating is the principal effect when short wavelengths are used.

Where susceptibility is high, as with the resonant condition described above, SAR correlates well with power density and can be an

[15]The axis length is approximating that of a resonant half-wave dipole.

[16]This relationship foretells the frequency-dependent nature of the guidelines for human exposure to RF energy.

[17]This depends partly on the dimensions of the body, since a very high length-to-width ratio could mean that the surface absorption is still affecting the major portion of the body's volume.

effective assessment parameter. As discussed later, both the *dosage* and the *dosage rate* can be important when assessing, and predicting, biological effects.

Weak levels of nonionizing electromagnetic radiation exist in virtually all accessible environments. For instance, a "strong" radio signal, one that the FCC requires its licensees to use for serving their principal communities, is 5 millivolts per meter (5 mV/m) at AM radio frequencies, 3.16 mV/m at FM radio frequencies, and 10 mV/m for UHF television frequencies. The corresponding "exposure" levels are 0.0000000066 mW/cm^2, 0.0000000026 mW/cm^2, and 0.000000027 mW/cm^2, respectively.[18] Radio stations are heard at even greater distances, of course, with their signal levels falling by orders of magnitude below even these low levels.

Biological Effects

The biological effects of RF exposure have been studied for decades. A number of summaries have been compiled, as individual research projects and as part of the review conducted by standard-setting bodies. These summaries do not necessarily involve new research; rather, all the data reported by other researchers is evaluated for quality and relevance, combined and then analyzed together, presumably with the data having been normalized to the extent possible to account for differences in experimental conditions. Such analysis is a useful tool when searching for dependence on one or two variables, such as frequency and SAR.

All of the many standards-setting bodies have followed such an approach. Discussed in more detail in Chap. 2, these analyses all find the presence of a similar threshold level, above which (about 4 to 5 W/kg SAR) adverse effects are reported but below which (about 3 to 4 W/kg) no such effects are reported.

Applicability of research

While extensively studied from many aspects over many decades, the entire body of research does not form a seamless cover for the whole range of possible conditions, nor does it provide a straightforward application to the establishment of human exposure guidelines. The conclusions that have been drawn from this work in setting guidelines, therefore, are intentionally conservative.

---

[18]While the prevailing standards limiting human exposure to RF energy are described in Chap. 2, it is worth noting here that these power flux densities are *millions* of times less than the standards allow for continuous exposure.

Virtually all of the early work was performed at 2,450 GHz, one of the most readily available of the industrial, scientific, and medical (ISM) frequencies. Later work has involved testing at 900 and 150 MHz, and there have been limited experiments performed at frequencies throughout the RF band. There has also been a wide range in the nature of the RF energy experimentally applied. It has been pulsed (as with radar), continuous (as with communications), and burst (as with a discharge), and it has often been amplitude modulated (as with AM radio and TV). These are all important attributes of RF energy, with potentially important implications for biological effects, but researchers have made independent decisions, of course, as to their own experimental variables.

Many researchers have performed experiments using subhuman mammals as test subjects. Most of these have been performed at very high levels of dosage, since that was generally what was required to find an effect. Less research has been done, until recent years, at low dose levels, since negative results were neither as interesting nor as likely to be rewarded with additional research funding.

It has been difficult to derive human exposure limits from such work, much less limits that apply over conditions very different from those of the specific experiment, in which RF energy may have been applied until some event (behavior interruption, convulsion, or death) occurred. One of the biggest difficulties is in extrapolating animal data to humans. Since *in vivo* experiments deal with organisms designed by nature to maintain their own temperatures within certain limits, it is vital to know whether the capabilities of that design are being exceeded by the experimental parameters. Thus, power density and duration must have been both controlled precisely and recorded precisely. Otherwise, the principal cause of the effects observed may simply be heat stress, which was induced by the RF energy absorbed, rather than the RF itself. Significant variations in core, brain, and extremity temperatures occur with physical activity, for instance, and in moderate heat stress, without indications that detrimental or lasting effects occur. Lower-order mammals may not have **thermoregulatory**[19] systems as adaptive or sophisticated; rats, for instance, are a laboratory animal of some convenience, but they are fur-bearing and do not sweat. Extrapolation of animal data to the human condition is problematic, especially when behavioral responses had been disabled (*i.e.,* the subject was drugged and/or restrained).

---

[19]Thermoregulation is the ability of a living body to maintain a characteristic or "set" temperature under a variety of environmental conditions.

The limited applicability of individual studies needs to be acknowledged, as well. Scientific research involves numerous variables, and isolating the one independent variable in experimental trials requires considerable effort. The hallmark of sound scientific inquiry is repeatability; unfortunately, the popular press does not recognize this. Within the scientific community, a single finding is not accorded much credibility unless and until it has been corroborated by other researchers in other laboratories. This is due to the complex nature of experiments that are looking for subtle changes; the process of measuring a phenomenon may affect its magnitude or even its existence, and the variables isolated may not be the controlling variables in nonexperimental conditions. For instance, RF at the exposure levels in most of these experiments does cause heating, and the subject animals are not unaware that they are being affected. As heating occurs, the thermoregulatory capabilities of the animals would be activated, including movement out of the RF fields. The inadvertent presence of visible or audible cues correlated to the application of RF energy may also generate learned responses, leading to escape from uncomfortable conditions. These are several reasons why anesthetization or restraint is normally utilized.

Another important element for accuracy in animal studies, one that has too often been ignored, is handling control groups as nearly identically as possible to test groups, even going so far as to "expose" them in an identical manner but to a zero field, since the mere handling of the subject animals adds stress, as does their possible restraint or anesthetization. Many of the reported studies fail in this simple regard, although most recent studies do include such "sham" protocols for the control populations.

Another category of RF exposure research is known as **epidemiological,** which is defined as the study of epidemics or epidemic diseases. Epidemiological studies, while retrospective, have the advantage of assessing human effects directly, so that findings might be applied directly and with more confidence than from animal studies. Such research with respect to RF exposure generally consists of statistical analysis of reported diseases and cross-referencing to presumed levels of RF exposure. In occupational settings, job classifications and military postings have been used as a proxy for probable levels of exposure, and hospitalization records and personal questionnaires are used to determine health effects.[20] In residential settings, zip code di-

[20]Chapter 2 contains a discussion of the epidemiological study of U.S. Embassy personnel subjected to chronic, low-level microwave irradiation in Moscow by the Russian hosts.

rectories have been used, for instance, to define areas around broadcast sites, and disease occurrence patterns by residence location are then evaluated, but the limited data on length of residency and the inability to control for other environmental conditions, especially during times spent out of the residence, have been complicating factors. Nevertheless, it is principally because of the paucity of observed effects from common levels of RF exposure that such studies have rarely identified even the *prospect* of any correlation, and those few that did have been refuted by subsequent attempts to duplicate their findings.

Many of the studies often mentioned as examples of the detrimental effects of RF exposure at levels complying with the prevailing standard (PS) fail to meet one or more of the simple scientific standards mentioned above, or the findings may have uncertain applicability to the human condition. Nevertheless, with the wide variation in experimental and study parameters and with the continuing research interest (see Chap. 5) that this field generates, the one overall observation that *can* be made is the absence of reported health effects at exposure levels below some threshold. As the National Council on Radiation Protection and Measurements stated in 1986,

> In spite of marked differences in field parameters, thresholds of behavioral impairment were found within a relatively narrow range of whole-body-averaged SARs ranging from ~3 to ~9 W/kg....Thresholds of disruption of primate behavior were invariably above 3 to 4 W/kg, the latter of which has been taken in this report, as well as by ANSI[-82], as the working threshold for untoward effects in human beings.

And as the American National Standards Institute stated in 1992,

> [T]he [body of research] literature is still supportive of the 4 W/kg criterion.... No verified reports exist of injury to human beings or of adverse effects on the health of human beings who have been exposed to electromagnetic fields within the limits of frequency and SAR specified by previous ANSI standards, including ANSI C95.1-1982.

Heat stress

Electromagnetic energy deposition creates heat within biological matter by the friction on adjacent particles from induced movement of individually charged molecules, atoms, and atomic particles. There are three modes of such movement, the development of which is dependent on the nature of the biological matter and on the frequency of the energy. First is the rotation of bipolar molecules to accommodate their preference to align themselves with the field; since the field is reversing polarity, the molecules try to flip back and forth to maintain the minimum energy configuration. Second is the reciprocal lateral motion induced in free electrons or highly charged atoms and molecules as the reciprocating field is applied. And third is the simple vibration induced in certain large molecules.

RF energy absorbed by a living subject is rapidly converted to thermal energy, which is then rapidly diffused within that animal by conduction with adjacent and/or deeper tissue and by convection due to blood flow; dissipation due to radiation is generally insignificant, although, in certain species, heat loss through respiration and perspiration can be a significant factor in heat dissipation.

Heat stress can be defined as energy deposition greater than the **thermoregulatory capacity** of the subject animal (which might be termed **SAR_{TRC}**). All animals generate heat naturally when stored energy from food is consumed as fat or sugars; this is known as the animal's **basal metabolic rate (BMR).** Thus, the SAR_{TRC} for that species will be at least high enough to accommodate that amount of energy; otherwise, the animal could not survive its own unlimited temperature increase. The BMR in humans has been estimated at about 1 W/kg, with exercise-induced metabolic rates of about 3 W/kg for light manual work, 6 W/kg for heavy manual work, and 10 W/kg during short, strenuous activities. Thus, SAR_{TRC} for humans is probably about 4 W/kg.

Heat stress will cause the body temperature of the test animal to increase so long as the energy continues to be absorbed. Nature has designed the animal to accommodate some temperature increase in elastic fashion, such that the animal returns to its preexposure state without change after exposure to the factor has ceased. Beyond that elasticity limit, which may be depend on a variety of other factors such as hydration, ambient temperature, and anxiety level, the animal may not return to normal.

Unfortunately, essential information on dose and rate of dose was often not recorded or not reported for experimental studies. In order to evaluate whether the reported effects from studies including that data can be considered "nonthermal," a **thermal stress factor (TSF)** can be defined, expressed in °C and given by

$$\text{TSF} = 0.236 \text{ cal/joule} \times 0.001 \text{ kg/g} \times (\text{SAR} - \text{SAR}_{\text{TRC}}) \text{ W/kg} \times \text{duration seconds}$$

TSF thus relates to the total dose (rate × time) of RF energy and can be determined from the SAR and duration of the experimental procedure. A negative TSF would mean that the animal had remaining thermoregulatory capacity and was not overly stressed; therefore, any observed, nonbehavioral effects could, indeed, be attributed to a nonthermal interaction. Positive TSF values, on the other hand, would indicate that the animal had been stressed thermally and that the principal cause of the effects observed may simply be heat stress, which was induced by the RF energy absorbed, rather than by the RF itself.

Light manual work, even for periods as long as 30 min, is not thought to raise human core body temperature more than 1°C. The National Institute for Occupational Safety and Health suggested in 1972 that 1°C was a reasonable upper limit of human tolerance, although it has been reported since that daily temperature variations of 1.5°C are normal and that transitory temperature increases of as much as 2°C can result from emotional stress.

Values of TSF greater than 1 to 2°C are understood to exceed the elasticity limit in many of the common animal subjects (rats, mice, rabbits, cats, dogs, and monkeys), which might have a lower SAR_{TRC} than humans. Such values can be lethal. If so, experimental exposures at low SAR, perhaps even below the human guidelines in the PS, may still be inducing thermal stress. In virtually all of the studies reviewed below, heat stress is a complicating factor, if not the causative factor.

Localized heat stress

The discussion above relating to thermal stress dealt with whole-body assessment. In situations involving exposures near to the RF source or involving high energy levels for short periods of time, localized areas of an irradiated body may be subjected to heat stress, while the energy deposition averaged over the entire body mass may have low TSF values or even negative ones. Since the thermoregulatory capability of a body incorporates several organic functions integrated together, not all of those may be available to accommodate localized thermal loads. Therefore, the threshold for observable or irreversible changes may be lower.

There are only two specific body parts considered to have limited thermal-stabilizing capabilities available, due principally to relatively low quantities of blood flow and to the self-contained nature of the structure. The eye mass is distinct from nearby muscle tissue and bone, and the ocular fluid is wholly contained within the eye organ. Therefore, heat cannot be carried away by convection of a common blood supply, and a significant temperature gradient may result. Similarly, the testes in males are also relatively isolated from the body, and the sensitivity of testicular function to heat is well known.

For other parts of the body, it appears that localized application of RF energy can be accommodated at higher levels than might be allowed for whole-body irradiation, due to physiological changes that serve to enhance blood flow, and thus heat convection, through that area. This is the principle behind therapeutic application of RF (see discussion below).

Specific concerns

Many specific concerns have been raised in the media, often highlighting tentative findings from a single research study. A number of these, if not a majority, allege the existence of nonthermal effects. While it is true that the effects observed are not strictly thermal, the conditions under which the effects were created almost always involve very high power density levels, leading to the possibility that the effects are, in fact, thermal injuries.

Cataract development. The oft-expressed concern about RF inducing cataract formation is a prime example of the generalization above. Rabbits were used in many of the animal experiments, and they had to be anesthetized or restrained in order to sustain the exposure levels applied to them and cooling jackets had to be supplied so that the exposures would not be lethal. Power densities above 100 mW/cm^2, and at durations of an hour or longer, were required to cause the development of cataracts. This gives specific absorption rates well in excess of the PS guidelines, and it appears that even higher levels are required in higher species in order to develop cataracts. This high amount of energy deposition severely stresses the animals' thermoregulatory systems, and the development of cataracts could be considered a thermal injury. Studies have shown that there is a threshold level of SAR, below which absorbed RF energy can be accommodated by the subject animal and will not lead to cataract formation. This indicates that there is no cumulative effect of the RF exposures, only damage when the localized thermal load exceeds the subjects' ability to dissipate it.

Studies have been made of cataract development in humans, focusing largely on military personnel, especially radar workers. Characterized by difficulties with sample size, extraneous factors, and control groups, such studies have not shown a correlation between RF exposure and cataract development at power densities below 100 mW/cm^2.

Blood-brain barrier permeability. The effect reported in 1977 of a temporary change in the permeability of small inert molecules across the blood-brain barrier of rats generated considerable media coverage at the time. Two studies, in 1978 and 1979, found similar changes only if the power density levels were high enough to cause a large increase in brain temperature, thus suggesting that this effect, too, is a thermal condition.

Calcium efflux. Much study has been made of the flow of calcium ions across the membrane of brain cells. Most of this work has been done on rabbits and cats, using RF field intensities ranging from many times the PS down to levels approximately equal to the PS. It is vital to note that the procedure used in these studies involved "preloading" the brain stem of the subject animal with positively charged calcium molecules and then observing the rate at which the molecules flow out (*efflux*) into the blood stream. Because of the preloading, there will be calcium efflux *whether or not* the animal is exposed to RF; the experiment was designed to record only changes in *rate* of efflux. Thus, it is incorrect to state that RF causes efflux of calcium ions from brain tissue, as even a body as august as American National Standards Institute has unwittingly done. None of these studies purports to show that.

Behavioral changes. In these experiments, rats, rhesus monkeys, or squirrel monkeys are trained to perform certain tasks, and then they are exposed to increasing levels of RF energy until the performance of those tasks is disrupted. Threshold power density levels for behavioral changes have been reported from 5 to 50 mW/cm^2, depending on the duration of exposure and the behavioral changes observed. Work stoppage and attempted movement out of the field are not necessarily indicative of a detrimental or permanent effect. Such behavioral changes are, in fact, an effective, integral part of the subject animals' thermoregulatory systems, and they can be expected to be evidenced at relatively low core temperature increases, giving rise to their sometime designation as "nonthermal" effects.

Chromosomal effects. Certain changes in plant and animal genetic codes have been reported from exposure of *in vitro* biological cultures to RF energy. While not necessarily exhibiting a significant overall temperature increase, thermal stress may still be the causal factor in light of the proclivity for localized intense thermal gradients to develop and the limited ability of a tissue *sample* to effect thermoregulatory actions of the complete organism from which the sample was taken. There is no corroborated evidence of genetic effects at low or even moderate power densities. Studies involving *in vivo* samples of rats indicate that power densities below 10 mW/cm^2 do not cause genetic changes.

Retarded growth. The inducement of abnormalities in offspring (known as teratogenesis, literally, the creation of monsters) has been widely performed by researchers, using mammals and nonmammals.

The power density levels used to effect such birth defects, and often the death of mother and/or offspring, have been obscenely high. Thermal concentrations and/or increases in core temperatures were expected and recorded as part of these experiments, which indicate very little of value about exposure to low or moderate levels of RF energy, except to note that near-lethal levels of exposure are often necessary to create such effects.

Immunology. Many of the early studies done of effects of RF exposure on blood composition were done in Eastern Europe, and many studies suffer from the structural shortcomings identified earlier. In addition, most findings of immune function suppression under RF exposure also involve power density levels sufficient and sometimes acknowledged to produce significant heating of and in the subject animal. Much of the speculation regarding possible compromise of immune functions centers on changes, either up or down, in the quantity of leukocytes. These are one of several types of blood cells, formed by various glands in the body, that help to maintain immunity against bacteria and other foreign bodies. Some studies took blood samples before and after irradiation, while others introduced foreign bodies into the animals' blood and monitored the systemic response.

Very few studies report on the impact of RF exposure to immune systems in intact animals at power density levels within the animals' thermoregulatory capabilities. Those that do, perhaps as one exposure level among several, have shown no or only limited effects.

Central nervous system. Most of the studies on laboratory animals have seen little or no effect on brain/nerve function for absorption rates below about 3 W/kg. Positive findings below that level have generally not been duplicated or corroborated. At levels above that, behavioral changes were observed, such as reduced sleeping time or changes in escape avoidance. Various markers[21] have been used to assess changes in brain activity, some of which require sacrifice and autopsy for evaluation, with few changes recorded, whether fields are pulsed or continuous.

The electroencephalogram (EEG) is often used to assess condition or changes in brain activity. Electrochemical impulses from about 1 to 100 Hz at amplitudes of about 1 to 10 V/m are used by the body for communication and coordination among various parts, organs, and processes. Such low-frequency signals are orders of magnitude below

[21]Common chemical markers are ATP (adenosine triphosphate), CP (creatine phosphate), and NADH (nicotinamide adenine dinucleotide).

the range of radio frequencies, and no direct interaction between the two would be expected, although modulation of RF radiation at frequencies in this range might affect brain/nerve function. Many of the early studies, however, suffered calibration difficulties with signals induced on metal electrodes and wire leads in high RF fields. More recent studies have not observed EEG changes at absorption rates below 4 W/kg.

Auditory effects. There is a recognized phenomenon known as microwave hearing. This occurs with modulated RF irradiation at frequencies from about 200 MHz to 3 GHz, and the person "hears" clicks or chirps located within, immediately above, or immediately behind the head. The mechanism is reportedly thermoelastic expansion within the cranium, detected by the ear, when the modulating frequency is in the audible range, via bone conduction. Depending on frequency and pulse width, average power density levels above about 5 mW/cm^2 are required to cause the effect. This same effect has been demonstrated to occur in animal subjects.

Anecdotal evidence

Of the many stories that are told of harmful effects of microwave radiation, there are three that are repeated so often that inclusion here is warranted. One is the case of actual injury to an employee at Sandia Corporation. This technician operated a microwave test apparatus and made a practice of holding his hand in the microwave beam and observing the presence of a heating effect in order to establish that the equipment was operational. He apparently also looked directly into the beam in order to place his hand in the right location, subjecting his head to power densities calculated at 100 mW/cm^2. After almost a year of such practice, he noticed a rapid loss of vision in a matter of days and was found to have developed cataracts.

Another involves the reputed chronic failure of a certain radar station on Friday nights. The story is that a radar technician was able to divert the microwave radiation into an adjacent room, entrance to which he allowed on Friday nights before he and the other staff at the site began their weekend leaves. The purpose of the exposure was to induce temporary sterility, for which he was able to charge considerable fees.

A third case involves the death of a woman in Florida in 1993 from a brain tumor. The growth was located in the area of her brain above her ear, and it was claimed that it was induced by the hours of use that the woman had made of her cellular telephone. This case was subsequently dismissed (see Chap. 4).

Low-level exposures

Thermoregulatory behavior can be initiated at levels of exposure too low to cause thermal stress. Nerves in the skin are able to respond to minute temperature changes (hundredths of a degree), and low-intensity RF fields can initiate thermoregulatory behavior in the same manner as other sources of heat. In addition, new findings have been reported recently of effects from exposure to RF at continuous levels of SAR so low that thermal stress is probably not the primary factor. However, even here it is not clear that such effects can be attributed to RF energy. An analysis reported in 1993 by Robert K. Adair of the interaction of RF radiation with small biological particles at the cellular level indicates that, even for power densities as high as 10 mW/cm^2, "the interaction of electromagnetic field with elements holding permanent charges or charge distributions will be masked by thermal noise and, hence, cannot be expected to generate biological effects."

Medical applications

Hyperthermia is a state of tissue temperature well above normal[22] induced by external forces, generally with therapeutic intent.[23] This is distinguished from a true **fever,** which is an internally induced temperature elevation due to an increase in the "set" temperature of the body, while hyperthermia is induced in spite of the existing set temperature. Hyperthermia resulting from RF energy absorption differs greatly from that occurring as a result of fever, resembling more closely hyperthermia induced by exercise.

The application of outside agents, including RF energy, is used to heat certain parts inside the human body for therapeutic purposes. The first reported use of RF dates from 1892, when it was observed[24] that RF could be passed through the body to provide the desired heating effect without the complications known to result from use of low-frequency currents. While there can be considerable variation in medical equipment, **diathermy** (also called **ultrasound**) uses contact applicators and frequencies up to about 1 MHz, while **radiothermy** (also called short-wave diathermy) uses noncontact plates at frequencies up to about 100 MHz. So widespread have such medical applications been that the FCC has set aside a number of frequencies for

[22]At or above 41°C (106°F) for humans, versus a normal body temperature of about 37°C (98.6°F).

[23]The quotation "Give me power to produce fever and I will cure all disease" is attributed to Hippocrates (*c.* 460–360 B.C.), a Greek physician.

[24]Credited to Jacques A. D'Arsonval (1851–1940), a French physicist.

their use. Diathermy has been and is still used with varying degrees of success for treatment, often in conjunction with other factors, of cancer and other disorders. Microwave radiothermy, the diathermic use of RF above 1 GHz, is not common due to reduced tissue penetration depth, greater variability in absorptive properties of body components, and excessive heating near the skin.

Radio frequencies can also be used for biomedical imaging. In X-ray imaging, the beam is well formed, and shadows from denser materials are relatively clear. At the longer wavelength of nonionizing frequencies, scattering from discontinuities in the body makes a less clear image, and corrective techniques are needed to enhance resolution.

Modeling of irradiated subjects

Extensive research work has been done with models of human beings and animals in work that is parallel to or complementary with experimentation on live test animals. The modeling work has been of two basic types: calculation and physical. With the availability of computing capacity to academic researchers, increasingly complex computer models of biological structures and bodies have been constructed. Using existing or refined algorithms for energy transfer on surfaces and through space, the temperature effects of irradiation with RF can be calculated with great precision. Physical modeling of biological structures and bodies has grown from leather bags filled with saline solution to highly complex arrangements of parts simulating skin, fat, muscle, blood, ligaments, and bone. These models can be irradiated in much the same manner as live test subjects.

Modeling research is popular because of several advantages over experimentation with live animals, advantages which pertain to both types of modeling. For instance, it is much easier to observe temperature changes *inside* a model, especially in real time for observation of transient effects. It is easier to rule out complicating factors, such as learned responses or unrelated health problems, and it allows real-time monitoring of internal temperatures. Modeling also should be faster, cleaner, and cheaper.

Unfortunately, models have only limited application. As largely static models, they cannot incorporate the range of dynamic responses in a living being, especially behavioral responses, but also physiological ones. For example, thermographic studies generally predict high localized fields at smaller cross-section locations, such as the neck, wrist, ankle, or tail, as well as at certain locations within the cranium, dependent on frequency of the applied energy. Such concentrations are often found not to occur in live subjects, however, presumably due to increased convective heat transfer by increased blood flow.

Models are also incapable of providing information on morphological changes that precede some irreversible biological endpoint, such as organ failure, cataract development, teratogenesis, or death. The thermoregulatory capabilities of living animals are truly remarkable, and there is no substitute for their direct observation.

Summary with emphasis on development of standards

Many demonstrated, measurable responses to RF irradiation remain within the range of the subject's physiological compensatory mechanisms and thus are not hazardous. A fundamental question in hazard assessment is whether to take the most sensitive function that could be taken as the determinant of a hazard level or whether to use an endpoint signifying a deleterious change in function. That is, not all effects are necessarily hazardous; in fact, as noted in the section above on medical applications, some effects can be beneficial.

Clearly, the biological effects of RF exposure are dependent on the specific frequency of the energy source. Difficulties are encountered in trying to scale the effects observed in small animals to those that might be encountered by human beings, and scaling of frequency must be considered, as well, particularly when the biological interaction involves whole-body resonance. Amplitude must be considered for scaling, too, since given RF field intensities will raise the body temperature of a small animal more quickly than that of a large one, due to its higher ratio of surface area to mass. Large bodies have a smaller relative surface area through which RF energy can be absorbed into their greater mass.

Further complicating the scaling of animal results to humans is the basic physical and physiological dissimilarity between different species. Similar biological processes in different animals are not necessarily affected to the same degree or even in the same manner by the same dose of RF energy or by the same absorption rate. Often, the test animals were selected for the ease with which the tests could be performed on them, rather than for the ease with which the findings could be extrapolated to the human condition.

There are three conclusions about human RF exposure that can be drawn with some certainty:

1. The human body is most susceptible to EMF energy absorption at about 80 to 120 MHz, due to long-axis resonance for bodies about 4 to 6 ft in length.

2. At frequencies above about 30 GHz, the depth of penetration into tissue is limited and energy deposition in large bodies is only superficial. At 30 GHz, for instance, penetration depth is only about

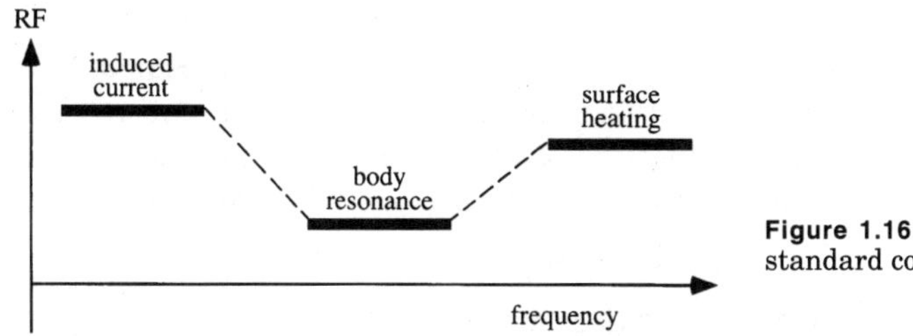

Figure 1.16 Basic exposure standard configuration.

½ inch (½ in) and so only about 1 percent of the total body mass is subject to exposure. Therefore, localized SAR, instead of whole-body SAR, becomes more important, and findings from infrared and/or laser research may be pertinent, as well.

3. At frequencies below about 3 MHz, near-field conditions apply and whole-body SAR will be low, with electric currents flowing through the body becoming the predominate factor.

Based on these conclusions, the basic exposure standard can be expected to look something like that shown in Fig. 1.16. Note the tightest restrictions occurring at whole-body resonance, due to the most effective coupling of RF to the human body at those frequencies; the relaxation at higher frequencies, where penetration of RF energy is not deep and the principal effect is surface heating; and the even greater relaxation at lower RF frequencies, where the human body is not resonant and absorption of RF energy is not very efficient. Dashed lines represent possible transitions from one realm to the next, although the nature of such transitions (*e.g.*, stepped, sloped, curved, or asymptotic) will be somewhat arbitrary.

The absolute levels of the limits will be based upon two figures: 1) SAR_{TRC} in humans, and 2) desired safety factor. The expression of those limits in terms of measurable parameters (*i.e.*, electric and magnetic field strength) will be based upon assumptions as to human interaction with those fields. Chapter 2 summarizes the efforts of standard-setting bodies to establish these figures and shows the progression of their assumptions over time as more research data has become available.

Intentional Radiators

Since the turn of the century, we have constructed an enormous variety of devices that radiate RF energy. (Those that do so *un*intentionally are discussed in the next section.) Those intentional sources can be grouped into two broad classifications: continuous and intermittent, with the latter further grouped into broad range and short range.

Due to historical patterns even predating the FCC's creation, and due to variations in the suitability of certain frequencies for specific purposes, there are scattered across the RF spectrum small segments assigned for military use and for categories of civilian use such as:

Public safety communications (*e.g.*, police and fire)

Government communications

Amateur communications

Land mobile communications (including ambulance)

Maritime communications

Aeronautical communications

Maritime navigation

Aeronautical navigation

Educational and entertainment broadcast

Common carrier (*e.g.*, telephone backbone)

Weather monitoring (via radar)

Satellite communications

Continuous radiators

The most significant intentional sources of RF are broadcast antennas, designed to radiate an RF signal on which data has been encoded with some modulation scheme. **FM radio stations** are licensed by the FCC at powers up to (and exceeding, in a few instances) 100 kilowatts (100 kW) effective radiated power, and **VHF TV stations** are licensed at powers up to 316 kW peak visual effective radiated power (or about 160 kW *average* power, with the inclusion of the aural signal). These stations, FM and VHF TV, operate between 50 and 220 MHz, in the frequency range at which most exposure guidelines are tightest.

The FCC authorizes **AM radio stations** at up to 50 kW nominal input power, which can become, with a directional antenna array, effective radiated powers of up to several hundred kilowatts, and the FCC licenses **UHF TV stations** at peak visual powers of 5,010 kW (or about 2,500 kW average effective radiated power). At AM and UHF frequencies, most exposure guidelines do rise severalfold, so these higher-power facilities are not necessarily of greater concern than are FM and VHF TV stations. Figure 1.17 lists the U.S. broadcast frequencies, along with maximum power levels in each band.

band	frequency range	maximum power
AM radio	535–1705 kHz	50 kW input
FM radio	88–108 MHz	100 kW radiated
VHF TV	54–88 MHz 174–216 MHz	100 kW radiated 316 kW radiated
UHF TV	470–806 MHz	5,010 kW radiated
Advanced TV	470–806 MHz	not yet set by FCC
wireless cable	2.150–2.162 GHz 2.500–2.686 GHz	8 kW radiated 8 kW radiated

Figure 1.17 U.S. broadcast bands.

The term **microwave** typically refers to radio frequencies above about 1 GHz, and microwave stations are licensed on frequencies as high as about 40 GHz. They generally do not send energy over a large area. Instead, this nonbroadcast service uses large "dish" reflector antennas to focus the energy tightly and direct it toward a similar receiving dish (this is known as *point-to-point* service). High radiated powers can be generated at microwave transmitting antennas, but exposures would be of concern generally only to those persons in close proximity to the antennas as a matter of occupation.

Microwave stations are also used to send signals to satellites, for retransmission to satellite receiving antennas. Figure 1.18 lists the designations for various microwave bands. While the receiving antennas do not generate RF, obviously the *uplinks,* or *earth stations,* do. The dish sizes are quite large, in order to provide sufficient area to focus the beam tightly enough to reach the satellite. Power levels, as a consequence, can be significant once the beam is fully formed, although, because of the angle at which the dish is looking up, maximum power density will occur some distance above ground.

designation	frequency range
L	1 – 2 GHz
S	2 – 4
C	4 – 8
X	8 – 12
Ku	12 – 18
K	18 – 27
Ka	27 – 40
V	40 – 75
W	75 – 110
millimeter	110 – 300

Source: Ferrel G. Stremler, 1990

Figure 1.18 Microwave bands.

Broadcast and microwave stations generally operate continuously, although some stations will save power by ceasing operations during late night and early morning hours. Also operating continuously are stations providing navigational aid to ships and planes. The satellite **GPS**[25] and the coastal **LORAN**[26] services operate by sending signals from multiple transmitters with signal modulation that allows a receiver to accurately measure the distance to each transmitter. With three or more distance measurements, plus the known locations of the transmitters, geographical position can be determined anywhere on the earth. GPS transmitters are satellites in high earth orbit, and power levels are strong only in the inaccessible vicinity near the satellite. At ground level the GPS signals are extremely weak. LORAN uses ground-based VLF transmitters, with radiated power levels up to several thousand kilowatts, so power density levels may be significant in the immediate vicinity of a transmitter. Radio beacon systems used for ship and aircraft navigation allow a position to be determined by triangulation if three or more beacon signals can be received, too, but more often a single beacon is used to fly or steer a course directly to or from the transmitter. The area served by a beacon is relatively small, perhaps a hundred kilometers across, and the power levels are relatively low.

Also operating continuously are **radar**[27] installations for air traffic control, weather monitoring, and national defense. Radar operates by sending out a single pulse and then listening for a much longer time to see what is reflected back. This gives radar a very low **duty cycle,** although the pulses themselves may be so strong that time-averaged power levels can be significant. Radar systems make active use of the microwave bands shown in Fig. 1.18, as well as HF, for over-the-horizon detection, and VHF and UHF, for long-range surveillance. Air traffic control radar can be used for airport surveillance (ASR) and for air route surveillance (ARSR). Many aircraft are themselves equipped with radar units, as are all ocean-going ships. Weather radar is most often located in remote areas central to the continental United States, and it monitors storm centers and weather patterns.

Intermittent—broad range

Most other intentional sources operate only upon demand. For instance, a radio dispatch service such as might be used by a delivery or

---

[25] This term stands for Global Positioning System.

[26] This term stands for Long Range Navigation.

[27] The term radar is coined from radio detection and ranging.

taxicab company operates only when the base operator or one of the mobile units keys up[28] for a short period to give a spoken message. There is more than one transmitter involved, since each mobile unit has a separate transmitter, as well as the base unit. Thus, radio frequency radiation (RFR) concerns are raised at each location, although there are at least two mitigating factors inherent in such services. First, each party to the communication alternates speaking, with the transmitter on, and listening, with the transmitter off, and the short periods of transmission provide a relatively low duty cycle. Most exposure guidelines are concerned with exposure *averaged* over some period of time, such as 6 min, that is usually considerably longer than a single transmission might last. Second, the power levels in use are generally quite low. In recognition of this, current exposure guidelines exempt RF devices operating below 1 GHz with less than 7 W of transmitter power, on the rationale that it is not possible for the exposure levels from such equipment to exceed the guidelines. Such transmitters are not normally licensed as to location, so one has little way of knowing where they are, but the power density levels affecting anyone not actually operating the equipment would be negligible.

Cellular telephone systems also have a relatively low duty cycle. A base station (*cell site*) may have 10 to 20 transmitters feeding an antenna serving a particular geographical area, with the number of transmitters operating that relates to the number of calls in progress. In some systems, each transmitter will run only as much power as is necessary to maintain connection with the mobile unit. There will always be one channel (called a set-up channel) operating to track mobile units entering or leaving the antenna's range should a call be placed to that unit while it is within range. Newer digital cellular technologies multiplex[29] several conversations onto a single transmitting frequency, which limits power variations and raises the duty cycle.

Another category of civilian use is **amateur (ham) radio.** Amateurs may operate, subject to FCC rules, in 26 reserved bands ranging from 1.8 MHz up to 250 GHz.[30] Amateurs can sometimes operate with considerable power, depending on the band in use and the particular equipment installed. Because of the operator's familiarity with the

[28]This colloquial term refers to depressing a spring-loaded button or lever, normally mounted on the side of a handheld microphone. This activates the transmitter and quickly becomes an automatic action before speaking.

[29]Multiplex in this usage means to combine several signals onto a single carrier, generally by time-sharing that carrier.

[30]Amateur transmissions above 300 GHz are authorized, as well, but this exceeds the frequency range normally defined as "radio."

equipment, and because of a typical intermittent and low duty cycle, most amateur use is exempt from the prevailing exposure standards.

Citizens band (CB) radio is distinct from amateur in that no licensing is required, although it is also for private use. Allowed power levels are much lower, and communication distances are much shorter. **Family Radio Service (FRS)** and **General Mobile Radio Service (GMRS)** are similar services more recently defined by the FCC.

Intermittent—short range

Short-range RF devices abound. These are units designed to radiate RF, but they operate with low power and hence are used for localized purposes. Within the household, one can find nursery monitors, cordless telephones, remote appliance switching,[31] intercom systems,[32] cordless microphones, and home security systems, among others. Out in the car, one can find gate and garage door openers, remote entry systems, and vehicle tracking transmitters. For private cars, the latter would be activated only in certain circumstances, such as theft of the vehicle in which it is installed. On the way to work, one might pass highway patrol using radar to check the speed of passing cars.[33] At work, vehicle tracking systems may be used, for instance, throughout their routes. Inside the factory, security systems use RF, as may real-time inventory systems.[34] On the way home again, one might find UHF bar-code scanners in use at the convenience store. The power on all these short-range devices is generally so low that exposure conditions cannot exceed the standards.

Modulation and multiplexing

On virtually all intentionally radiating RF systems there is a need to encode data onto the RF wave (the exception being security systems, where interruption of the beam is all that is necessary to trigger the alarm). This encoding is done by varying the **continuous wave (CW)** at the assigned radio frequency from the transmitter in some manner that is recognizable by the receiver. With AM radio, for instance, the encoding is done by **amplitude modulation (AM),** meaning that the relative strength of the signal varies, and the receiver is

---

[31]Most TV and cable remote controls use infrared frequencies.

[32]Some intercom systems work over in-house ac wiring.

[33]An active jammer in the private car would, of course, be illegal.

[34]Some inventory control systems record all their data for later transfer to a computer.

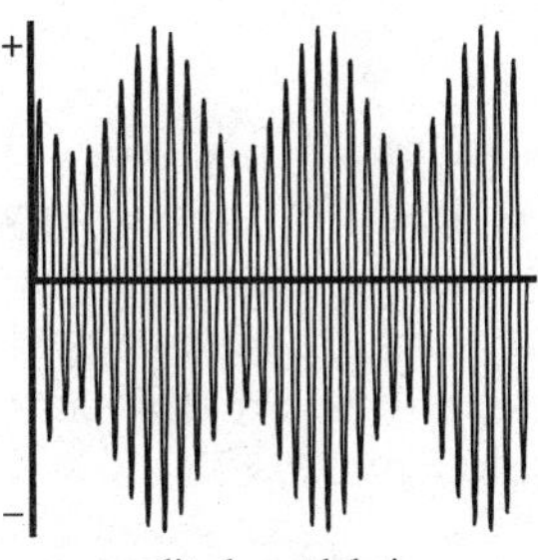

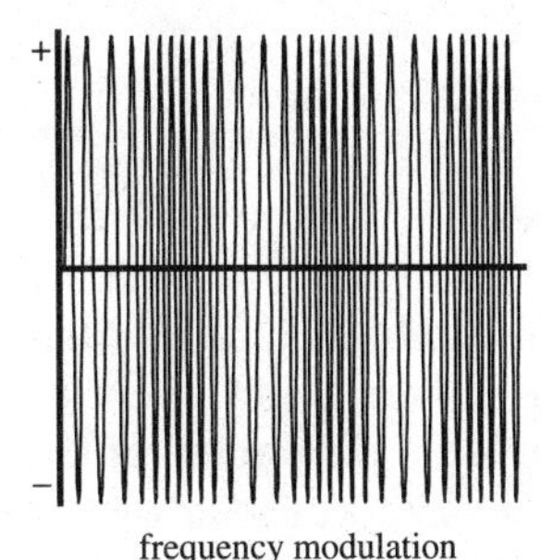

Figure 1.19 Modulation types.

able to decode those variations and construct an analog signal to drive acoustic speakers. For FM radio, the encoding is done by **frequency modulation (FM),** meaning that the frequency is varied just a little from the center frequency assigned to that station, and the receiver is able to decode those variations and drive speakers. Figure 1.19 shows examples of amplitude and frequency modulation. Television uses a complex modulation scheme, with its video carrier being amplitude modulated and its aural carrier being frequency modulated.

First-generation cellular systems were all analog FM systems, with the human voice directly modulating the carrier frequency of the transmitter. One RF channel is required for each voice signal. The use of multiple RF frequencies to provide simultaneous access to multiple subscribers is called **frequency division multiple access (FDMA).**

Second-generation cellular systems employ **digital modulation** in which the voice broken into short samples that can be transmitted more efficiently. Because of this, it becomes possible to transmit *several* voice signals in the space of *one* RF channel. One way of combining the voice signals is by transmitting a small portion of each in turn. Assuming eight voice signals, they could be transmitted sequentially like this:

```
1 2 3 4 5 6 7 8 1 2 3 4 5 6 7 8 1 2 3 4 5 6 7 8 1 2 ...
```

This technique is referred to as **time division multiple access (TDMA).** In this example, if there were fewer than eight "calls," say, just one, it would not be necessary to transmit during all eight time slots. In fact, cellular TDMA base stations *do* transmit continuously; they fill in the unused slots. The subscriber telephones, on the other hand, transmit *only* during their assigned slot, like this:

```
1               1               1               1  ...
```

Code division multiple access (CDMA) is another scheme for digitally combining several calls onto one channel.

The new **Personal Communications Services (PCS)** are all digital systems, and various of the large PCS spectrum auction winners are committed to one or another of these technologies, including **Groupe Spéciale Mobile (GSM),** which is a European cellular TDMA standard. Delivery and performance problems with some of the newer schemes should greatly affect the relative success of the providers. Claims of differences in their potential to interfere with pacemakers is discussed in Chapter 5.

In the realm of digital television, there are also several competing data encoding schemes: **QAM,**[35] multiple level **VSB,**[36] and **OFDM.**[37] Because this industry is not being driven by consumer demand, a single standard will probably be adopted before major financial commitments are made.

Gain

The antennas used for radiating the signals are designed to focus their energy toward the corresponding receivers. The receiving antennas may be fixed (rooftop TV antenna, for instance) or mobile (car radio), but they are generally located within a fairly narrow range of elevation angle and orientation from the transmitting site. Therefore, the system engineer does not want the transmitting antenna to send energy either up or down, where there are no receivers, but tries to send it out toward the horizon.[38] The degree of this focusing is known as the **gain** of the antenna. A hypothetical point source (like a tiny light bulb)

---

[35]Quadrature amplitude modulation.

[36]Vestigial sideband modulation.

[37]Orthogonal frequency divisional multiplexing.

[38]Except, of course, in the case of satellite uplinks.

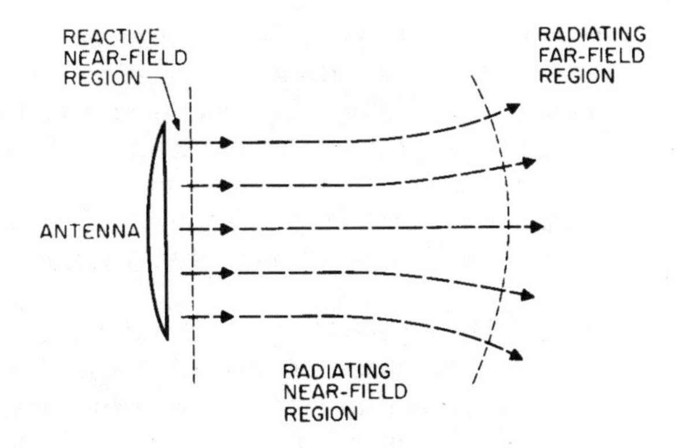

Source: Richard C. Johnson, 1984.

Figure 1.20 Antenna near- and far-fields.

would radiate equally in all directions, and sometimes the gain of an antenna is expressed in **dBi,** or dB relative to such an isotropic radiator. A half-wave dipole antenna is easily formed from two wires, each a quarter of a wavelength long at the frequency of interest; its performance has been well characterized, and often the gain of some other antenna will be expressed in **dBd,** or dB relative to a dipole.

Considerable focusing can occur, especially with dish antennas, and peak powers in the main direction of the antenna can be significantly higher than the power fed to the antenna. Of course, power is not *generated* by the antenna, only focused, so a high gain in one direction implies that there are lower gains in other directions. It takes some distance in front of the antenna for the focusing to occur. The area farther from the antenna than this distance is known as the **far field,** while the area closer to the antenna is known as the **near field,** as shown in Fig. 1.20. The demarcation between the near and far fields depends on the physical size of the antenna and the frequency being radiated, but rules of thumb for the distance to the far field include 10 wavelengths or 10 times the height of the antenna (see Chap. 3).

Propagation

Radio signals at different frequencies propagate with different characteristics, which help determine siting considerations for different services. The higher frequencies, especially microwave, above about 1 GHz, have wavelengths so small (no larger than a foot) that they cannot bend around buildings or hills. It is said that such services need **line of sight** in order to propagate, and so such transmitting facilities are typically sited on hilltops or tall buildings and towers. The receiving antennas for microwave service typically need to be outdoors so that they have line of sight to the transmitting antenna. This factor also affects UHF TV (frequencies above 470 MHz), where outdoor antennas are needed unless field strengths are high enough to allow building penetration of sufficient signal. To a lesser degree, VHF TV and FM (wavelengths no longer than 20 ft) are affected by the same line-of-sight limitation, in that their signals will not reliably reach beyond hills or mountains.

AM, by contrast, has a relatively long wavelength (a tenth to three-tenths of a mile) that can bend around obstacles without excessive attenuation. Of course, with wavelengths that long, the area surrounding an AM antenna is particularly important to forming its radiating wave. For this reason, AM stations are almost always sited in flat areas with, preferably, no other radiating structures nearby.

Incidental Radiators

Natural sources

Nature's use of EMF ranges from the massive to the minuscule. The principal natural source of EMF on earth is, of course, the sun. That enormous furnace is slowly consuming itself while radiating energy at all frequencies across the spectrum. Of most importance for life on earth are the components of heat and light. The speed at which the earth rotates on its axis allows fairly even heating of its surface, which the atmosphere partly retains, and life, including human life, has flourished in several successive cycles as a result.

Animal life is designed by nature to utilize light to trigger biological processes and for certain photochemical reactions to take place in the skin organ. Plants are designed to utilize light energy via photosynthesis for growth and reproduction. It is unclear just what parts of the rest of the EMF spectrum, if any, are actively used by animals and plants.

The sun exhibits increases in radiation from time to time. Short-term increases of 1,000 times or more in radiation amplitude occur, varying in length from seconds to hours. Longer-term increases in solar radiation are known to correlate with the appearance of sunspots. RF radiation in the range 0.5 to 40 GHz is most affected, although the terrestrial impact of the sunspots includes short- and medium-wave (AM) radio transmission propagation, since sunspot radiation affects the degree of ionization within the layer of the atmosphere known as the ionosphere, which in turn affects the ability of various layers to absorb or reflect such radio signals back to earth.

The number of sunspots has been observed to vary on a cycle of approximately 11 years.

Other extraterrestrial sources of EMF energy are other stars and star systems. Very high energy radiation, including X radiation and cosmic rays, is studied by researchers to gain an understanding of the structure of the universe and the observable star systems within it, as well as its possible origin. The earth receives tiny amounts of energy from such sources continuously, and this includes RF components, although the earth's static magnetic field shields much of what would otherwise reach us at the surface, and the atmosphere filters out much of the rest. RF energy below about 10 MHz is largely absorbed by water and oxygen molecules, while EMF above about 40 GHz is reflected by the ionosphere, an interesting exception being visible light.

Organizations like Search for Extra-Terrestrial Intelligence (SETI) monitor the RF regions of the cosmic spectrum for signs of any regular patterns in the noise that might indicate the presence of intelligent life on other worlds. This program is based on the recognition that we have, for almost a century, been sending out such signals ourselves, advertising to the universe our presence on the third planet of an ordinary star on the edge of an average galaxy; as yet, there has been no response.

RF energy of modest intensity is generated on earth by atmospheric disturbances, such as thunderstorms, tornadoes, and hurricanes. There is wide variation in locale and season, of course, and the largest amplitude components are at frequencies of 2 to 30 kHz. The earth itself emits so-called black-body radiation due to its temperature above absolute zero; the total thermal noise is on the order of 0.0003 mW/cm^2 over the entire RF spectrum. The human body also emits black-body EMF at a similar rate and also generates EMF internally for biological feedback and other activities only now beginning to be understood.

Artificial sources

The ongoing manufacture and delivery of RF-generating products, and the ongoing introduction of new products using RF, creates a constantly rising low level of RF background noise. While natural sources of RF exist, as noted above, with few exceptions such sources are overwhelmed by artificial RF. Of course, the RF that is generated does not accumulate; there is no buildup of high RF concentrations anywhere. As already noted, all EMF radiates at the speed of light and so dissipates quickly.

Besides the intentional generators of RF energy listed in the preceding section, there are also artificial sources of *incidental* RF, for which exposure standards would also apply. This is radiation in the

RF spectrum from products whose principal purpose is the performance of some other task. Even the low levels of RF radiated from such products can far exceed the levels from natural sources, although this is only true at close range, since the natural sources are so diverse and at such a large distance that the resulting fields are almost uniform, while the power of the artificial field drops as the square of the distance from its source.[39]

The prime example is the now-common microwave oven. This product generates up to 1,500 W of RF energy at 2,450 MHz and focuses it inward. When introduced in the 1960s, the microwave oven was not targeted at the consumer market, but it did elicit attention from the government. In 1968 Congress passed the Radiation Control for Health and Safety Act, which authorized the Department of Health Education and Welfare and the Federal Drug Administration to enforce a maximum limit on microwave emissions of 1 mW/cm^2 at time of manufacture, as measured at 5 cm from the unit, and not more than 5 mW/cm^2 for the life of the unit. Other home uses of RF include ultrasonic cleansers and humidifiers.

There are several industrial applications that rely on the generation and application of RF, without the intent to radiate any RF beyond the item to which the RF is being applied. Heat *sealers,* or *welders,* can use RF; they operate by generating RF energy and letting RF current flow between two plates, as shown in Fig. 1.21. In between the plates can be placed plastic items, and the current will flow through the plastic, as well. When high levels of RF are generated for a short burst, the plastic is heated to its melting point, or at least to a point of malleability. With careful design of the apparatus and adjustment of the frequency, power, and duration, plastic joints can be made efficiently. Such heat sealers are used for making sandwich bags, air mattresses, and other products out of plastic. Many vinyl products are also formed, patterned, and joined in this manner, automotive parts being a common example.

[39] Recall the Inverse Square Law.

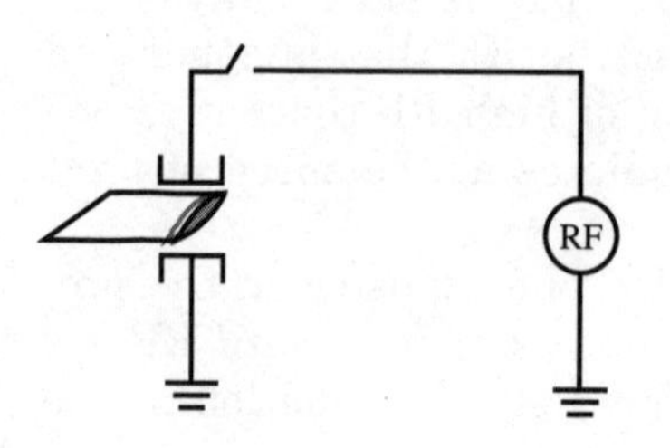

Figure 1.21 RF heat sealer schematic.

6.78 MHz	±0.015 MHz
13.56	±0.0071
27.12	±0.163
40.68	±0.02
915	±13
2.45 GHz	±0.05 GHz
5.8	±0.075
24.125	±0.125
61.25	±0.25
122.5	±0.5
245	±1

Note: Specific ISM frequencies in use worldwide vary from country to country.

Figure 1.22 U.S. ISM frequencies.

As discussed earlier, there are many medical applications for RF energy. These are not intentional radiators, because they are designed to focus the RF energy on the subject tissue, much as a heat sealer is designed to focus the RF on the plastic objects. The FCC has designated several frequencies throughout the RF spectrum for such purposes, these being the so-called **industrial-scientific-medical (ISM)** frequencies. Figure 1.22 lists the ISM frequencies in use in the United States; the specific frequencies assigned vary from region to region around the world.

In addition to heat sealers, RF is also used in industrial settings for such thermal applications as hardening of metals, welding, soldering, melting, refining, plating, production of material with special electrical properties (*e.g.*, semiconductors), drying and bonding of wood products, food processing, particle separation, disinfection, and insect control. In most ISM applications, there is little interest in careful control of frequency, spurious emissions, or even, to some degree, amplitude. Combined with cost constraints in the design and manufacture of this equipment, as distinguished from broadcast equipment, considerable leakage of RF can occur. Interference easily occurs between ISM machines, and interference to nearby electronics in some other application can also be common. Especially as new medical equipment is introduced and put into use, interference concerns become important. Many ISM devices have emission standards, promulgated by the FCC, the FDA, or some other agency, limiting radiation from the device generally to avoid **electromagnetic interference (EMI).** Occasional enforcement of such standards has occurred, but that task has become overwhelming and response is only made to specific complaints.

With regard to public exposure, leakage from most ISM equipment is not serious, although occupational exposure levels from certain high-power industrial RF devices can be considerable. Concern is especially warranted in such situations because workers at these machines can be in close proximity to them for long periods of time, and the issues raised in Chap. 3 on implementation of the RF exposure standards are particularly relevant.

Video display units (VDUs), also known as **video display terminals (VDTs),** are now ubiquitous in residential as well as commercial settings. They are also incidental sources of RF radiation, due either to the scan rate of the monitor screen or to the clock rate of the computer itself. While typical maximum fields do not exceed about 15 V/m or 0.08 amperes per meter (0.08 A/m) at 30 cm from the screen, there are several vendors selling screen covers designed to limit the levels of radiation from video monitors.

Noise

In addition to artificial sources of incidental RF radiation from devices that at least use the RF in some fashion, there are also artificial sources of RF that do not intend to make RF energy at all. Such sources of RF noise include power-line discharges (below about 50 MHz), electric trolleys and buses (below about 50 MHz), automobile ignition systems (up to about 155 MHz), and fluorescent lights (microwave). The first two can be annoying sources of interference to reception of TV Channels 2, 3, and 4, but none of them is much of an issue with respect to RF exposure conditions.

Chapter

2

History of Regulation

Early Standards

Before 1960

High-power microwave transmitting equipment was developed in the 1940s and 1950s by the U.S. military and by private companies, including Bell Telephone (Bell), American Telephone and Telegraph (AT&T), and General Electric (GE). As the operating powers grew higher, exposures of workers grew more frequent and more intense, and anecdotal evidence of microwave injuries became more commonly mentioned (see Chap. 1).

In 1953, the Central Safety Committee of the Bell Telephone Laboratories issued a bulletin that cited an example of visual acuity reduction at a power density of 100 mW/cm^2 and that recommended a 30 dB safety margin, leading to a recommendation of 0.1 mW/cm^2 for use within the company. This is believed to have been the first human exposure guideline promulgated in the United States. In 1954, General Electric Health Services recommended an upper limit of 1 mW/cm^2 for the exposure of its employees to microwave radiation.

In 1955, the Mayo Clinic in New York held a seminar on this topic, with papers given by American industry, the U.S. Air Force, and foreign researchers, as well. At the time, microwave diathermy procedures (see Chap. 1) were regularly used at the Mayo Clinic. One of the papers was presented by Lockheed Aircraft Corporation, which reported on a several-year research project with human subjects. The Lockheed finding was that no serious effects were recorded for exposure conditions as high as 13 mW/cm^2.

The U.S. military, meanwhile, was also developing its first standards, although public acknowledgment was not always made concurrently. In 1955, the U.S. Air Force apparently adopted an upper limit of 10 mW/cm^2; this information was presented at the first "Tri-Service" Conference[1] held in 1957 at the Rome Air Development Center, near Rome, New York. Two more "annual" conferences of this type were held at Rome in 1958 and 1959, with the published proceedings forming the first summary of research in this field.

Also in 1957, Bell and AT&T in coordination adopted 10 mW/cm^2 as the upper limit for their companies, with 1 mW/cm^2 as the limit for continuous exposure. GE adopted a 10 mW/cm^2 limit in 1958, as did the U.S. Army. In 1959, a "General Conference on Standardization in the Field of Radio-Frequency Radiation Hazards" was held at the American Standards Association (ASA) in New York. It was recommended at that time that the ASA establish a standardization project, reviewing the available studies from all sources and recommending a national standard.

All of these studies had been performed at microwave frequencies, and their range of applicability was generally 30 MHz and above.[2] Figure 2.1 shows the Bell Telephone Standard adopted in 1957. It is

[1]Representatives from the U.S. Army, Navy, and Air Force attended, with contractual assistance from George Washington University.

[2]Mega*cycles,* in those days.

1. For the time being, microwave exposure limits may be classified as follows:

Average Power Density	Classification
> 10 mW/cm^2	Potentially hazardous
1–10	Safe for incidental or occasional exposure
< 1	Safe for indefinitely prolonged exposure or permanent assignment

2. Employees are cautioned to abide by the following rules:
   a) Never enter an area posted for microwave radiation hazard without verifying that all transmitters have been turned off and will not be turned on again without ample notice.
   b) Never look into an open waveguide which is connected to energized transmitters.
   c) Never climb poles, tower or other structures into a region of possible high radar field without verifying that all transmitters have been turned off.

Figure 2.1 1957 Bell Telephone recommendations.

impressive, indeed, that in less than 4 years Bell and others progressed from recognizing a need to developing, critiquing, and adopting a standard, one that is not dissimilar from what we have today, almost 40 years (and hundreds of studies) later. For instance, the 1957 Bell Standard sets a basic limit of 1 mW/cm^2 for exposures of unlimited duration, which is the way the current standards express their basic limit. Further, the 1957 Bell Standard allows higher exposure levels for shorter time periods; again, this is the way that the current standards account for the human body's ability to accommodate a dose *rate* above equilibrium if the total dose does not exceed reversible levels.

Department of Defense

In 1965, the U.S. Army and Air Force amended their use of the prevailing 10 mW/cm^2 exposure guideline to include a time limit for exposures, given by the formula

$$T = \frac{6000}{S^2}$$

where S is the power density (mW/cm^2) of exposure and T is the maximum recommended exposure duration, in minutes. Thus, for S = 1 mW/cm^2, T = 100 hours (100 h), which is well beyond any shift length, and thus that would be allowed, essentially, for exposures of unlimited duration. For S = 10 mW/cm^2, though, the limit allowed by the standard then prevailing, the allowed exposure time was T = 1 h, which represented a significant effective tightening of the standard.

USAS C95.1-1966

The American Standards Association complied with the request at the 1959 General Conference by initiating in 1960 a Radiation Hazards Standards project. The project was taken under the joint sponsorship of the U.S. Department of the Navy and the Institute of Electrical and Electronics Engineers (IEEE), and ASA defined the scope as follows:

> Hazards to mankind, volatile materials and explosive devices which are created by manmade source of electromagnetic radiation. The frequency range of interest extends presently from 10 kHz to 100 GHz. It is not intended to include infrared, X-rays, or other ionizing radiation.

By the time the Navy and IEEE were ready to recommend a standard, in 1966, ASA had become the United States of America Standards Institute (USASI). That year, USASI adopted and published USAS C95.1-1966, "Safety Level of Electromagnetic Radiation with Respect to Personnel." The applicable frequency range was 10 MHz to 100 GHz, and it set a limit of 10 mW/cm^2 for unlimited exposures, with higher exposures allowed so long as any 0.1 hour average exposure did not exceed the basic 10 mW/cm^2. While this Standard represented a tenfold relaxation of the basic 1 mW/cm^2 limit in the 1957 Bell Standard, it equaled the Army, Navy, and Air Force limit, as well as that adopted by GE.

C95.1-1966 is quaint in some ways and prescient in others. Only three terms were defined: whole body irradiation, partial body irradiation, and radiation protection guide. The latter was defined as, "Radiation level which should not be exceeded without careful consideration of the reasons for doing so." (In contrast, C95.1-1992 defines 32 terms, taking more pages for that purpose than the text for the entire C95.1-1966.)

The 1966 Standard disclaims that its exposure guide "may be subject to revision as more knowledge is gained," but its basic finding, that "[r]adiation characterized by a power level tenfold smaller will not result in any noticable [sic] effect on mankind," has not yet been contradicted. Importantly, it introduced the $\frac{1}{10}$ h (6 min) period for time-averaging exposure levels; that time period would be utilized by virtually every Western standard promulgated since.

ANSI C95.1-1974

By 1974, USASI had become the American National Standards Institute (ANSI), which it remains today. Again under the sponsorship of the Navy and IEEE, ANSI adopted C95.1-1974, intending to incorporate only minor revisions to C95.1-1966. Actually, the wording changes appear to remove the time-averaging provisions for continuous wave radiation, allowing time-averaging only for modulated fields, perhaps meant to apply to radar, with its marked difference in peak and average power levels. It is presumed that this was inadvertent since no discussion of this change is contained C95.1-1974 itself. It does indicate, however, how seemingly harmless wording choices can have dramatic implications, an issue that would become more of a problem as the C95.1 Standards became more complex.

ANSI C95.1-1982

Background

After the ANSI C95.1-1974 Standard was published, active research work continued, especially by many of the principals involved.

Subcommittee IV[3] of ANSI C95, on Safety Levels and/or Tolerances with Respect to Personnel, met with some regularity. An important summary of the Subcommittee's deliberations was made by Dr. Arthur W. Guy, Chairman of SC-4, and published in the *Proceedings of the Non-Ionizing Radiation Symposium* held by the ACGIH in Washington, D.C., on November 26–28, 1979.

In preparing to adopt revisions to ANSI-74, SC-4 made four basic findings:

1. "No verified reports exist of injury to or adverse effects on the health of human beings who have been exposed to [radio frequency electromagnetic] fields within the limits of frequency and power density specified in previous ANSI guides."
2. Physical and biological data is "quite limited," and "previous guides were based on the assumption that only gross thermal effects...are potential causes of biological reactivity." The subcommittee expressly wanted to avoid "prejudgment in light of unsettled questions of field-body mechanisms of interaction and of emerging data that indicate the existence of athermal effects."
3. "[P]revious ANSI guides have been interpreted widely as occupational standards....The subcommittee recognized the need for a general-population guide."
4. ANSI-74 was in wide use, and its withdrawal "was considered highly undesirable." However, "[r]etention of the [flat, 10 mW/cm^2] 1974 guide was also viewed as undesirable in light of new data and developments," so it was decided to issue a revised standard "in spite of acknowledged gaps that persist in the existing base of data."

The revised standard developed from these findings was approved by ANSI on July 30, 1982.

Revised standard

Thus, as ANSI-82 itself modestly notes, "[t]he 1982 Radio Frequency Protection Guide (RFPG) is an extension of its 1974 predecessor with several notable refinements." The principal refinements are as follows:

1. *Recognition of whole-body resonance.* A frequency dependence was introduced, as shown in Fig. 2.2*a* and *b*, to reflect the fact that the human body absorbs energy much better when the wavelength

---

[3]This subcommittee is known as SC-4 ("S C four").

Frequency	Electromagnetic Fields		
Applicable Range (MHz)	Electric Field Squared (V^2/m^2)	Magnetic Field Squared (A^2/m^2)	Equivalent Far-Field Power Density (mW/cm^2)
0.3 – 3.0	400,000	2.5	100
3.0 – 30	4,000 x $(900/f^2)$	0.025 x $(900/f^2)$	$900/f^2$
30 – 300	4,000	0.025	1.0
300 – 1,500	4,000 x (f/300)	0.025 x (f/300)	f/300
1,500 – 100,000	20,000	0.125	5.0

Note: f is frequency of emission, in MHz.

Figure 2.2*a* Table of ANSI Standard C95.1–1982 Radio Frequency Protection Guide.

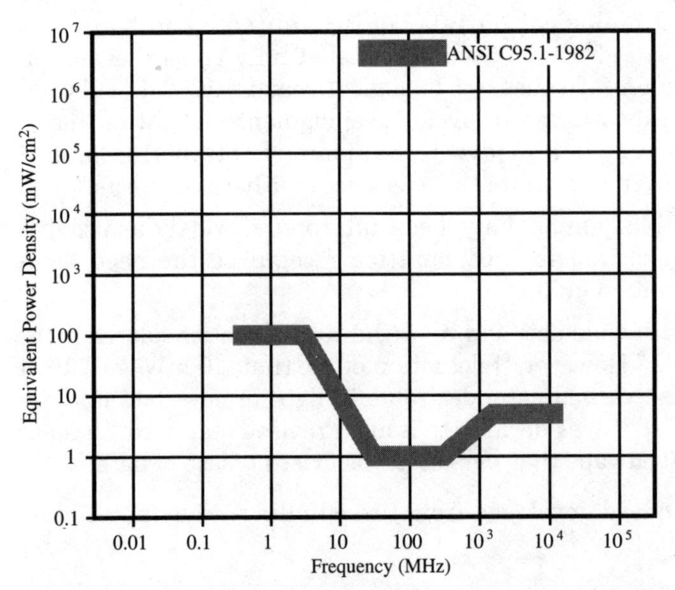

Figure 2.2*b* Graph of ANSI Standard C95.1–1982 Radio Frequency Protection Guide.

is about twice the body length (see Chap. 1). The frequency range of 30 to 300 MHz for the most restrictive portion of the new standard was selected to encompass the corresponding range of human lengths.

2. *Incorporation of dosimetry.* The underlying purpose of the guide is now to limit the specific absorption rate (SAR); the power density (or field strength) limits in which the guide is expressed have merely been judged adequate to limit SAR to the threshold levels.

3. *Broadened assessment criteria.* With the addition of new studies, "[t]he most sensitive measures of biological effects were found to be based on behavior....The whole-body-averaged SARs associated with thresholds of reversible behavior disruption were found to range narrowly between 4 and 8 W/kg....The subcommittee's intent was that of protecting exposed human beings from harm by *any* mechanism, including those arising from excessive elevations of body temperature [emphasis in original]."
4. *Safety factor.* The subcommittee noted that "reliable evidence of hazardous effects is associated with whole-body averaged SARs above 4 W/kg." This threshold would have been judged adequate for occupational exposures, but the subcommittee wanted to incorporate "a considerable degree of conservatism...in the RFPG to make it applicable to the control of nonoccupational as well as occupational exposures." Therefore, an order of magnitude safety factor (*i.e.,* 10 times) was selected and the SAR threshold was lowered to 0.4 W/kg.

No change was made in the 6 min time-averaging provision.

FCC adoption

In final action on February 26, 1985, under its General Docket 79-144,[4] the FCC adopted ANSI C95.1-1982 as the standard that all FCC licensees would need to meet. The effective date for applying this standard was January 1, 1986, and all FCC licensees were to bring their operations into compliance by that date. In fact, there was considerable uncertainty about how ANSI-82 was to be applied, and most licensees took little action until the television license renewal cycle began on June 1, 1986. That was the filing deadline for Maryland, Virginia, West Virginia, and the District of Columbia, with the rest of the states following in 17 more quarterly groups through April 1, 1989; the radio station renewals lagged behind by 2 years. The FCC required each station, at the time it filed for license renewal, to certify specifically that it was in compliance with ANSI-82. Stations applying for any changes in facilities before renewal had to provide certification at that earlier date.

To assist broadcasters with the assessment of RF exposure conditions and with deciding on necessary steps to comply with the new requirements, the FCC Office of Science and Technology prepared

[4]The docket number prefix indicates that the FCC began the adoption process in 1979, even before the standard it would adopt, ANSI-82, was even released.

Bulletin No. 65 and released it in October 1985. This document, known as OST-65, provided excellent guidance to the industry. First, it presented a tutorial on RF exposure. Formulas were given for the calculation of RF power density levels for all types of stations (see Chap. 3). In particular, OST-65 recommended that a ground reflection factor of 1.6 be assumed for the purposes of calculation; this has been in near-universal use ever since, and experience has shown that it is generally conservative.

Second, OST-65 provided tables by which individual stations, if there were no others nearby, might determine whether ground-level compliance could be presumed.

Simply because the widespread application of RF exposure criteria to broadcast stations was so new, OST-65 was later found to have given inadequate treatment to several issues. Two of them, multiuser sites and hot spots, were clarified by FCC action in Docket 88-469. The industry has since developed solutions for one other, the consideration of on-tower exposures.

Docket 88-469

There was active disagreement within the broadcast community in 1986 about the specific terms of ANSI-82 and how compliance was to be demonstrated. An evening "workshop" on the topic at the 1986 National Association of Broadcasters Engineering Conference, scheduled for a small meeting room, was packed with concerned broadcasters and featured a lively discussion among the consulting engineers present. Some argued that hot spots did not need to be measured or reported, because they could be detuned by someone putting a hand on them. Some argued that *both* electric- and magnetic-field levels had to be measured and that *both* had to exceed ANSI-82 in order for the guideline to be exceeded.

The author's firm consistently applied the most conservative interpretations in the assessment of RF exposure conditions at clients' transmitting sites (see Chap. 3). On July 15, 1987, with the experience gained at many such sites, the author's firm filed a Request for Declaratory Ruling with the FCC, seeking (and suggesting) clarification of the FCC's interpretation of ANSI-82 and its application to the certifications by licensees. Three issues were raised, and suggested concepts were made for the resolution of each:

1. *The definition of a broadcast "site."* While OST-65 had directed that all stations at a site share responsibility for bringing a site into compliance, it seemed unreasonable that stations which might contribute only a small amount of RF to some other area of an an-

tenna farm should bear responsibility for RF exposure problems that might exist there. It was suggested that an area where a station contributes less than 5 percent of its ANSI limit be considered a separate site for compliance purposes.

2. *The definition of "significant."* Under the National Environmental Policy Act of 1968 (NEPA), the FCC had to consider whether an action it might take had a significant effect on the environment. Because so many broadcasters were not taking an active role to achieve compliance, stations wanting to construct new facilities at a site, or even stations wanting to make a change in their transmitting facilities that might improve the RF exposure situation, could not do so until all the stations were cooperating. It was suggested that 5 percent again be used, such that a station contributing less than 5 percent of its ANSI limit at a site could proceed without consideration of the other stations.
3. *The definition of "hot spots."* ANSI specified that measurements were to be made no closer than 5 cm from reradiating objects, but experience had shown that many hot spots at antenna farms extended as far as 20 cm (8 inches) before the localized fields dropped below the ANSI field limits. Because such hot spots were still so localized that whole-body-averaged exposure would probably not be affected, it was suggested that the minimum measurement distance be set at 20 cm (8 inches).

The FCC upgraded the Request for Declaratory Ruling to become Docket 88-469, and in action on December 20, 1989, accepted all three suggestions. The percentage specified for items 1 and 2 was set at 1 percent, which was tight but still very helpful, and the distance specified for item 3 was given as "10–20 cm." Since no guidance was given for what might justify the larger distance, 10 cm (3.9 in) had to become the new minimum distance; this, too, was helpful by removing the majority of hot spots from the need for specific mitigation.

NCRP

Background

The National Council on Radiation Protection and Measurements (NCRP) was chartered by Congress in 1964 as a nonprofit corporation with four tasks[5]:

---

[5]This standard description appears in all NCRP publications.

1. Collect, analyze, develop, and disseminate in the public interest information and recommendations about (*a*) protection against radiation and (*b*) radiation measurements, quantities, and units, particularly those concerned with radiation protection;
2. Provide a means by which organizations concerned with the scientific and related aspects of radiation protection and of the radiation quantities, units, and measurements may cooperate for effective utilization of their combined resources, and to stimulate the work of such organizations;
3. Develop basic concepts about the radiation quantities, units, and measurements, about the application of these concepts, and about radiation protection;
4. Cooperate with the International Commission on Radiological Protection, the International Commission on Radiation Units and Measurements, and other national and international organizations, governmental and private, concerned with radiation quantities, units, and measurements and with radiation protection.

NCRP is the successor to the unincorporated association of scientists known as the National Committee on Radiation Protection and Measurements and was formed to carry on the work begun by the Committee. Due to its creation by Congress, NCRP has been the body to which the Environmental Protection Agency has deferred for guidance in this field, following its own withdrawal. (See further discussion in subsequent section on EPA.)

NCRP is made up of the members and the participants who serve on the many scientific committees of the Council. The scientific committees, composed of experts having detailed knowledge and competence in the particular area of the Committee's interest, draft proposed recommendations. These are then submitted to the full membership of the NCRP for careful review and approval before being published. Thus, the main body of work by NCRP comprises a series of publications, sequentially numbered from its incarnation as a Committee, that deal with some aspect of the matters falling under its purview. There are three Reports of direct relevance to RF radiation.

Report No. 67 (1981). *Radiofrequency Electromagnetic Fields—Properties, Quantities and Units, Biophysical Interaction, and Measurements.* This publication was designed to provide standardization of units in the field of RF radiation, and it laid the foundation for later research and standard-setting activity. Of most significance was the introduction of the term **specific absorption rate,** or **SAR** (see Chap. 1). This alternative to the term **dose rate,** already in use for ionizing radiation, where cumulative dosage is significant, has become a standard unit of measure for RF radiation. The consistent use of SAR by ANSI-82 the following year greatly helped to popularize its use for nonionizing radiation. Because it is expressed per unit mass, it has simplified the task of extrapolating experimental findings across species of widely varying size.[6]

Report No. 86 (1986). *Biological Effects and Exposure Criteria for Radiofrequency Electromagnetic Fields.* This publication is the "NCRP Standard" to which reference is often made. It is primarily a thorough (and very readable) review of research to that time (1982 was the cutoff publication date for consideration of research studies). The exposure recommendations, discussed below, were a minor part of the 382-page document and were not necessarily well developed as a standard.[7]

Report No. 119 (1993). *A Practical Guide to the Determination of Human Exposure to Radiofrequency Fields.* This publication was designed to provide a guide for those persons newly responsible for the assessment of RF exposure conditions and who are not already familiar with the principles, equipment, or practice of this area of specialization.

Exposure guidelines

Report No. 86 begins with this assessment:

> The lack of quantitative data on the biological effects of RFEM fields[8] has resulted in widespread concern that such exposure poses the risk of injury to health regardless of intensity. Although there are several thousands of reports—scientific papers, books, articles, and newspaper accounts—of widely varying scientific quality that present data or opinion on the biological response to RFEM radiations, no consensus has emerged regarding thresholds and mechanisms of injury at specific absorption rates (SARs) below a few watts per kilogram (W/kg). The wide variation in RFEM-radiation exposure criteria around the world reflects this ab-

[6]This is also the rationale behind the introduction in this book of the term **thermal stress factor** (see Chap. 1), *i.e.,* to simplify the task of extrapolating experimental findings across species with widely varying thermoregulatory systems.

[7]The Report is not consistently definitive in the discussion of its recommended exposure limits. For instance, it discusses "limits of exposure below 30 MHz, *and perhaps at frequencies somewhat higher,* apply...." (emphasis added), and it casually mentions "For other conditions,...these limits should be lowered" without bothering to suggest by how much.

[8]NCRP's term "RFEM fields" is equivalent to the term "RF EMF" as used in this book.

sence of consensus. An objective analysis of the scientific literature and recommendations for exposure limits by a qualified and unbiased group of experts is sorely needed.

At the time (1986), this was true, although it is no longer. Certainly, Report No. 86 itself provided the objective analysis so "sorely needed." The Report begins its review of published studies at the cellular level, increasing the complexity of the biological systems under study and ending with studies of human beings. Its basic findings can be summarized in a few quotations, as follows:

1. "In the absence of human data, it is necessary to turn to data on subhuman species in full realization that body dimension and mass have an enormous controlling influence on the SAR at a given frequency."
2. "[B]ehavioral disruption appears to the most statistically significant end point that occurs at the lowest observed SAR."
3. "Thresholds of disruption of primate behavior were invariably above 3 to 4 W/kg, the latter of which has been taken in this report...as the working threshold for untoward effects in human beings in the frequency range from 3 MHz to 100 GHz."
4. "[A]n appropriate margin of safety...has been taken as a factor of 10 for occupational populations."
5. "[I]t is recommended that there be an average exposure criterion for the general public that is set at a level equal to one-fifth of that of occupationally exposed individuals."

Figure 2.3*a* shows the limits recommended by NCRP. Figure 2.3*b* shows the NCRP limits compared with the earlier ANSI-82 standard, and Fig. 2.3*c* shows NCRP compared with the later ANSI-92 limits.

Frequency	Electromagnetic Fields						Contact Currents
Applicable Range (MHz)	Electric Field Strength (V/m)		Magnetic Field Strength (A/m)		Equivalent Far-Field Power Density (mW/cm^2)		(mA)
0.3 – 1.34	614	*614*	1.63	*1.63*	100	*100*	200
1.34 – 3.0	614	*823.8/f*	1.63	*2.19/f*	100	$180/f^2$	200
3.0 – 30	1842/f	*823.8/f*	4.89/f	*2.19/f*	$900/f^2$	$180/f^2$	200
100 – 300	61.4	*27.5*	0.163	*0.0729*	1.0	*0.2*	no limit
300 – 1,500	$3.54\sqrt{f}$	$2.59\sqrt{f}$	$\sqrt{f}/106$	$\sqrt{f}/238$	f/300	*f/1500*	no limit
1,500 – 100,000	194	*106*	0.515	*0.23*	5	*1*	no limit

Note: f is frequency of emission, in MHz.
Public limit is shown in *italics*.

Figure 2.3*a* NCRP Report No. 86 (1986).

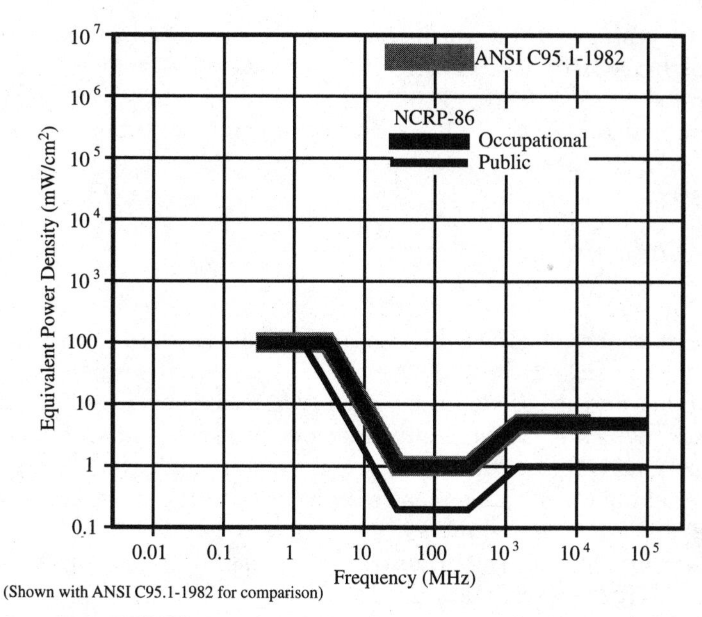

Figure 2.3*b* NCRP limits compared with the earlier ANSI-82 standard.

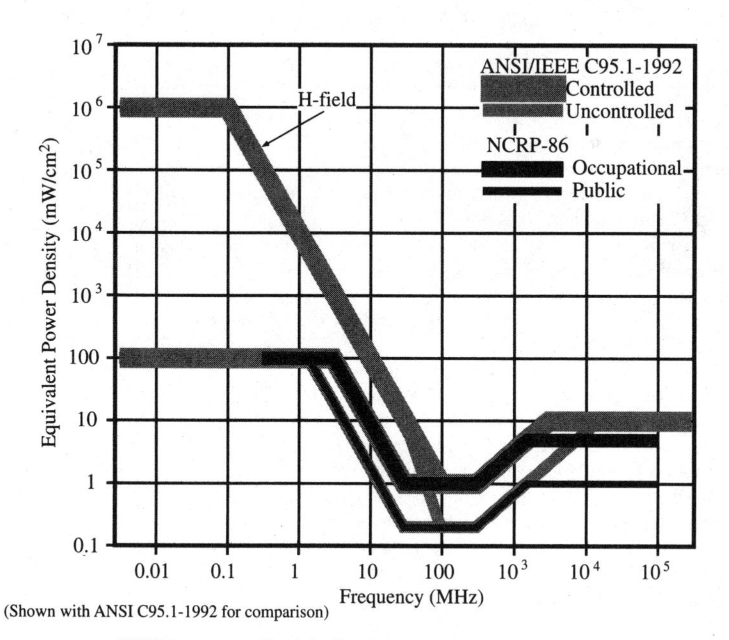

Figure 2.3*c* NCRP compared with the later ANSI-92 limits.

By now, every major standard in the world has followed the guidelines laid out by NCRP: 4 W/kg threshold, occupational safety factor of 10, and public safety factor of 50. Thus, as shown in subsequent chapters, the consensus that NCRP found missing in 1986 has, in large part due to NCRP's efforts, been achieved.

Foreign Standards

Just as the interest and concern regarding biological effects of RF radiation have spawned considerable research and standards development here in the United States, so, too, have they led to extensive study in Canada and overseas. Many countries have developed standards of their own, and several supranational agencies have developed standards, as well. In fact, there are more standards than can be included in a text such as this one, although many of the more significant ones are described below. Figure 2.4 is a partial list of the acronyms used by many of the organizations researching, promulgating, or enforcing RF exposure standards.

There is some variation among the standards that have been developed, due to combined effects of several factors: (1) difference in set of studies selected as relevant for setting a human exposure standard,

ACGIH: American Conference of Governmental Industrial Hygienists.
ANSI: American National Standards Institute.
CEC: Commission of the European Communities, Brussels.
CENELEC: European Committee for Electrotechnical Standardization, Brussels.
FCC: Federal Communications Commission.
ICNIRP: International Commission on Non-Ionizing Radiation Protection, varies.
IEEE: Institute of Electrical and Electronics Engineers.
ILO: International Labour Office, Geneva.
INIRC: International Non-Ionizing Radiation Committee (defunct), varied.
IRE: Institute of Radio Engineers
IRPA: International Radiation Protection Association, varies.
NATO: North American Treaty Organization, Washington, D.C.
NCRP: National Council on Radiation Protection and Measurements.
NIOSH: National Institute for Occupational Safety and Health.
NIRP: National Institute of Radiation Protection, Stockholm.
NRPB: National Radiological Protection Board, Great Britain.
OSHA: Occupational Safety and Health Administration.
OST-65: FCC Office of Science & Technology, Bulletin No. 65, October 1985.
STANAG: Standardization Agreement, NATO.
WHO: World Health Organization, Geneva.

Notes: All terms relate to American organizations unless otherwise noted.
Certain international organizations do not have fixed bases of operation.

Figure 2.4 Who's Who of RFR acronyms.

(2) differences in interpreting those studies and extrapolating to the human condition, (3) differences in evaluation of risk assessment and appropriate safety factors, (4) influence of preceding standards by the same or cooperating bodies, and finally, (5) differences in the intended applicability of the standard to subgroups of the nation's or region's population. What is remarkable, however, is the general agreement among all of these standards-setting bodies (1) about the underlying threshold for health effects, at 4 W/kg as averaged over the whole body, and (2) about a reasonable safety factor, at one order of magnitude. The difference among them seems to be more in how that safe level is to be recognized in measurable quantities.

NATO STANAG 2345

Prior to 1982, the U.S. Department of Defense, and the other nations forming the North Atlantic Treaty Organization (NATO), relied upon the frequency-independent 10 mW/cm^2 standard in general use, along with the U.S. Army and Air Force time limits discussed earlier. In that year, the Military Agency for Standardization promulgated the first version of NATO's own Standardization Agreement (STANAG).

The assumed threshold of adverse effects was found, yet again, to be an SAR of 4 W/kg, to which the familiar tenfold safety factor was applied, for an effective limit of 0.4 W/kg. Separate E-field, H-field, and power density limits were designated to approximate that SAR limit. Shown in Fig. 2.5, the 1982 NATO STANAG 2345, like ANSI-82,[9] recognized the frequency dependence of human absorption of RF energy, although it did not reflect the band of highest absorption at about 100 MHz.

In 1988, NATO's General Medical Working Party promulgated a revision to STANAG 2345. Its principal changes were a tightening from 30 to 100 MHz and a relaxation below about 10 MHz. An averaging time of 6 min was also incorporated, leaving the new STANAG looking very much like ANSI-82, as shown in Fig. 2.6.

WHO and IRPA

The Task Group on Radiofrequency and Microwaves, a joint enterprise between the **World Health Organization (WHO)** and the

[9]RF exposure standards are generally promulgated by agencies dealing with many factors affecting health and safety, and thus they may have rather lengthy titles to distinguish them. Nevertheless, most are referenced in this text by the format *body-year.* For instance, ANSI Standard C95.1-1982 is denoted ANSI-82, even though ANSI may have issued many other standards in that same year.

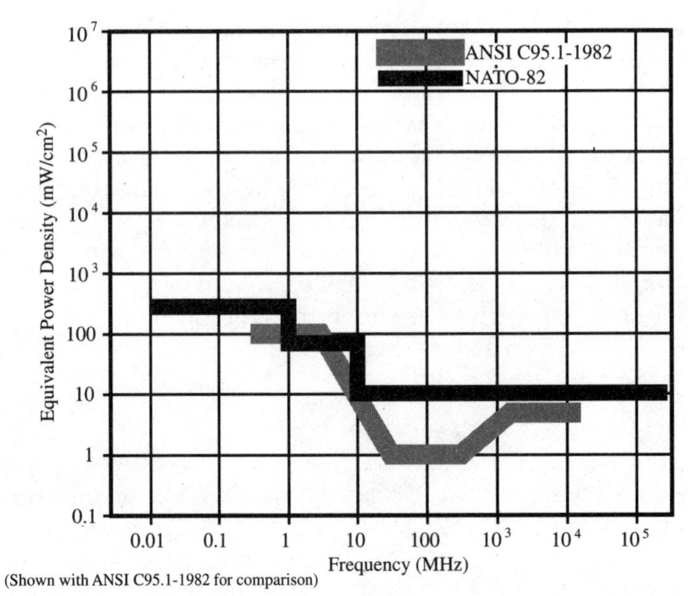

(Shown with ANSI C95.1-1982 for comparison)

Figure 2.5 NATO-82 RF exposure limits.

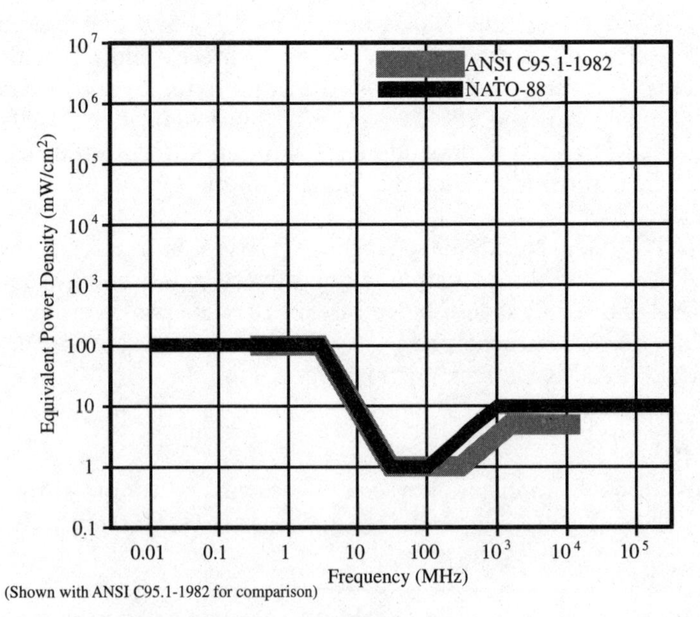

(Shown with ANSI C95.1-1982 for comparison)

Figure 2.6 NATO-88 RF exposure limits.

International Radiation Protection Association (IRPA) suggested in 1981 that power densities not exceeding 10 mW/cm^2 could be allowed for occupational exposure continuously throughout a working day. These guidelines were published by WHO as Environmental Health Criteria 16 (EHC16-1981). Lower limits were suggested for exposure by the general population, as well as higher limits for occasional exposures at frequencies in certain ranges, although neither of these adjustments was specifically detailed.

In 1984, IRPA issued more specific recommendations, based on a threshold of 0.4 W/kg for continuous exposure and setting forth basic limits in terms of electric and magnetic field strength. IRPA revised those recommendations in 1988, and a summary is shown in Fig. 2.7, along with ANSI-92 for reference. IRPA-88 is important because it represents the first adoption of two aspects that would become common among many later standards: (1) limits on the current flow in the body as induced by the surrounding fields, rather than by contact with a conducting surface, and (2) relaxation of magnetic field limits at lower frequencies.

The WHO published Environmental Health Criteria 137 in 1992, summarizing research findings, theories of biological interaction, and

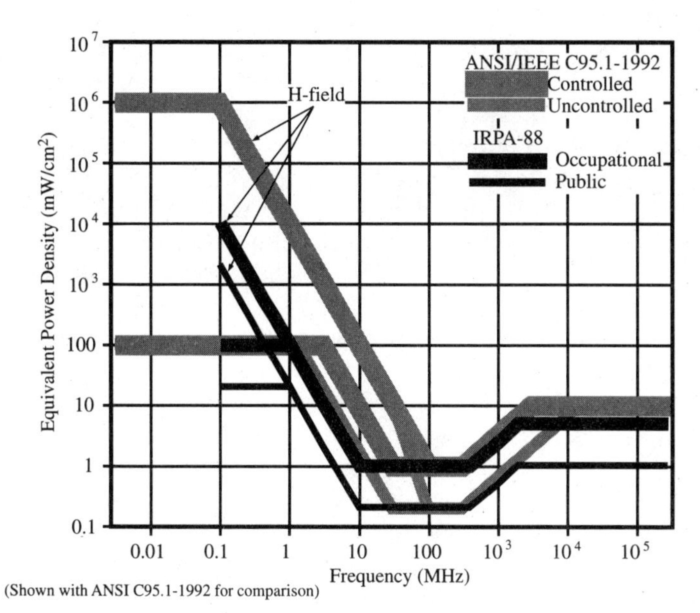

Figure 2.7 IRPA-88 RF exposure limits.

other issues pertinent to the regulation and mitigation of RF exposure conditions. The exposure standard described in EHC137-1992 is IRPA-88.

In May 1992, IRPA established the International Commission on Non-Ionizing Radiation Protection (ICNIRP) to continue the research, advice, and guidance that the predecessor International Non-Ionizing Radiation Committee (INIRC) of the IRPA had provided from its creation in 1977.

ILO

The **International Labour Office (ILO)** is an independent organization concerned with worker safety in many industries around the world. In 1985, the ILO published a general document on occupational hazards from nonionizing EMF, as well as a more specific technical review of RF exposures in 1986. The publications were intended for use by workers in affected industries and reflected the IRPA-84 Standard. In 1992, further guidance was issued, reflecting the IRPA-88 Standard.

Germany

Germany's development of RF exposure standards mirrors that in the United States. German standards have been voluntary, with the national government not having issued any standard. Prior to 1978, there was a prevailing 10 mW/cm^2 guideline in place, adopted by the German Association for Radar and Navigation. In 1978, Committee 764 of the German Electrotechnical Commission (VDE) published the first draft of an RF exposure standard, revising it in 1978 and again in 1984. In 1991, a significantly revised standard was issued, designated **DIN/VDE 0848,** with Part 2 of that standard setting limits for RF exposure.

While DIN/VDE 0848 does use a two-tier approach, it is unusual in that the public exposure criteria are more restrictive across the entire frequency band, actually becoming more so at the lower frequencies, instead of approaching the occupational limits, as do most other standards. Figure 2.8 shows the limits, expressed in equivalent far-field power density, compared with ANSI-92.

CEC

The **Commission of European Communities (CEC)** dates back as far as 1970 and has, as part of its Action Programme implementing

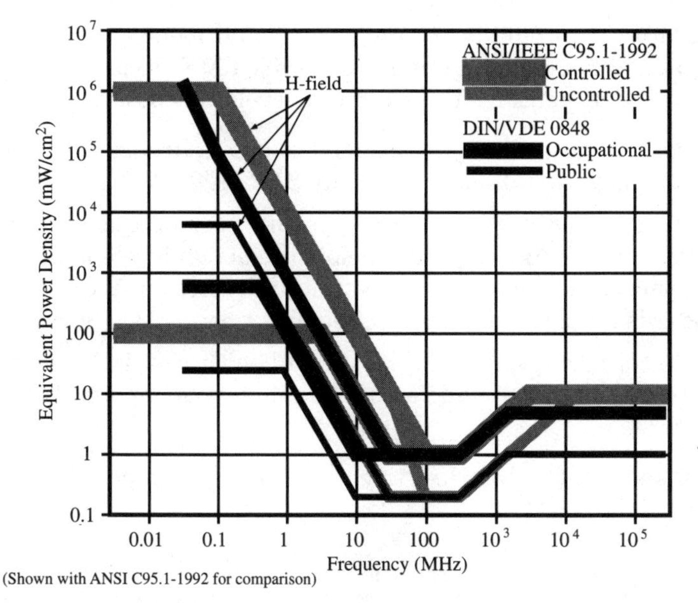

Figure 2.8 DIN/VDE 0848 RF exposure limits.

Austria	Germany	Netherlands
Belgium	Greece	Portugal
Denmark	Ireland	Spain
Finland	Italy	Sweden
France	Luxembourg	United Kingdom
As of 1995		

Figure 2.9 The European Community.

the Community Charter of Basic Social Rights for Workers, sought to develop consensus on RF exposure limitations among the various organizations with which it maintains a liaison. In 1991, the CEC proposed a very simple set of "Basic Restrictions for Workers" that acknowledged the common 0.4 W/kg safety standard threshold. While not providing a presumption for correlation to electric or magnetic field strength, the CEC-91 pronouncement is the second appearance of limits on induced body current.

There are 15 member states in the European Community organization, as listed in Fig. 2.9. Most have had some enabling national legislation to adopt some prevailing guideline, whether it is ANSI-82 or -92, IRPA-88, ACGIH, or in Britain's case, NRPB. The Commission desired to set a standard that would apply uniformly to all member states,[10] and in 1992 issued a Council Directive covering EMF from 0 to 300 GHz.[11] That Directive identified three discrete levels of exposure:

Ceiling levels are those which are deemed to pose a hazard of unacceptable risk. They are set, for whole-body SAR, at 0.4 W/kg.

Action levels are those above which exposure may be a cause for concern. This implies that exposure below the action levels is not a cause for concern, corresponding to the guidelines in ANSI and other standards for exposure of unlimited duration.

Threshold levels are those below which no health risk is present, and they are included to provide a goal for employers implementing the Directive.

Figure 2.10 shows the CEC Directive action levels, compared with

[10]The Commission also sought to produce a single document covering agents affecting human health grouped into four categories: noise, mechanical vibrations, EMF up to 300 GHz, and EMF above 300 GHz (optical radiation).

[11]In keeping with the focus of this text, subsequent discussion of the CEC Directive will be limited to the frequency range 10 kHz to 300 GHz.

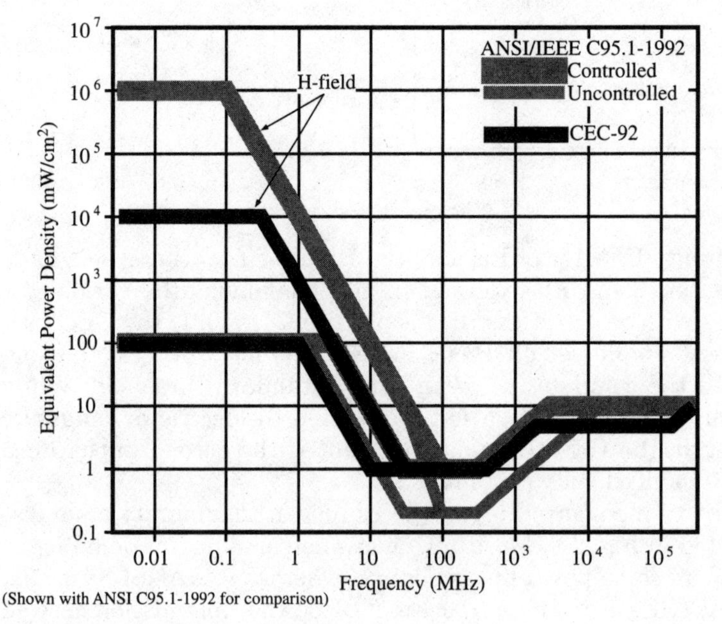

(Shown with ANSI C95.1-1992 for comparison)

Figure 2.10 CEC-92 RF exposure limits.

ANSI-92. Several subtle differences are evident, mostly as to frequencies for slope changes in the curves. CEC-92 represents the first occurrence of relaxed limits above 150 GHz.

United Kingdom

In 1960, the U.K. Post Office published "Safety Precautions Relating to Intense Radiofrequency Radiation," the first standard dealing with public exposure protection. It set a limit of 10 mW/cm^2 for exposures of continuous duration, regardless of frequency, and included practical guidance for ensuring compliance.

The **National Radiological Protection Board (NRPB)** was created in Great Britain in 1970 and has provided advice to the British Parliament since, conducting research in the field of protection against both ionizing and nonionizing EMF. In 1993, the NRPB published Vol. 4 No. 5 1993 of the Documents of the NRPB. This publication revised the 1989 report, NRPB-GS11, and takes a slightly different rhetorical approach to setting a standard. Rather than working from a target maximum SAR, such as 0.4 W/kg, and setting frequency-dependent field limits that are presumed to correspond to such an SAR, the NRPB simply set 0.4 W/kg as its "basic restriction" and then established "field investigation levels" for electric fields, magnetic fields, and contact currents. These field investigation levels can be exceeded if the basic restriction SAR is not, much as the ANSI, NCRP, and other field limits can be exceeded if it is shown that the underlying SAR threshold is not exceeded. The range of the NRPB standard includes extremely low frequency (ELF), but the RF investigation levels are as summarized in Fig. 2.11, which is shown relative to the ANSI-92 guidelines, for ease of reference. In several cases, as expected, the NRPB investigation levels are more restrictive than the ANSI maximum permissible exposures, although both are based on a basic restriction of 0.4 W/kg. The reason for this is the intent by NRPB to set limits based on the interaction of small children with the field. Thus, in order to demonstrate compliance, the NRPB standard is somewhat more burdensome.

Averaging times for NRPB-93 apply only to the basic SAR restriction, and they are 15 min for whole-body SAR and 6 min for partial-body exposures. The SAR for partial-body exposures is relaxed by a factor of 25 for head, fetus, neck, and trunk and 50 times for the limbs. Body currents are restricted only below 10 MHz.

A two-tier guideline is suggested by NRPB, as well, with its relaxation of the electric field investigation levels at frequencies above 10 MHz for areas "where it can be established that there is no possibility of small children being exposed." The relaxation in power densi-

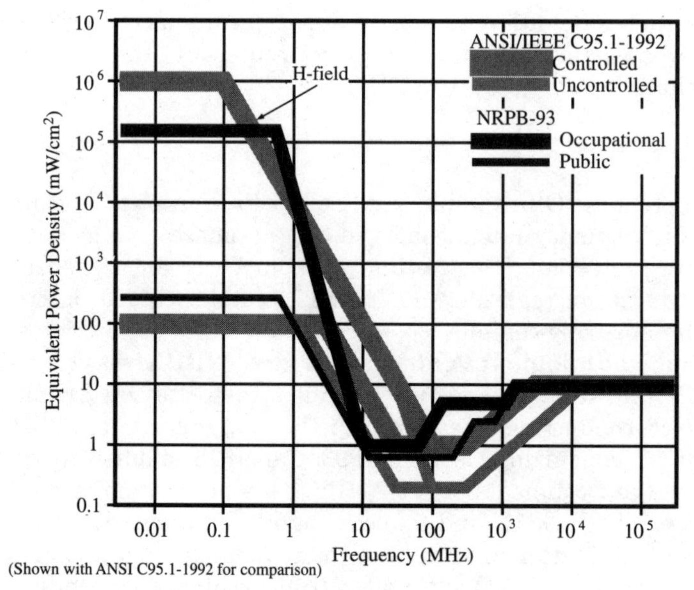

Figure 2.11 NRPB-93 RF exposure limits.

ty is by ratios up to 7.6, which allows adult-only exposures exceeding the ANSI-82 limits by up to 5 times.

CENELEC

The **European Committee for Electrotechnical Standardization (CENELEC)** covers 18 countries, the 15 European Community nations plus 3 European Free Trade Association (EFTA) countries. In September 1994, CENELEC issued a draft standard, styled as Prestandard No. 2. It was expected that all member nations would have adopted the Standard within 12 months through appropriate national legislation.

The CENELEC Standard uses specific reference levels for exposures of unlimited duration. The curve shape is somewhat simpler than CEC-92, but it does include the relaxation of limits above 150 GHz. Figure 2.12 shows this standard, compared with ANSI-92. The range of body current limits for CENELEC extends to 3 MHz, with guidance only to 100 MHz.

Canada

The Health and Welfare Ministry in Ottawa has adopted a variety of Safety Codes for protection of Canadian citizens. In 1979, the

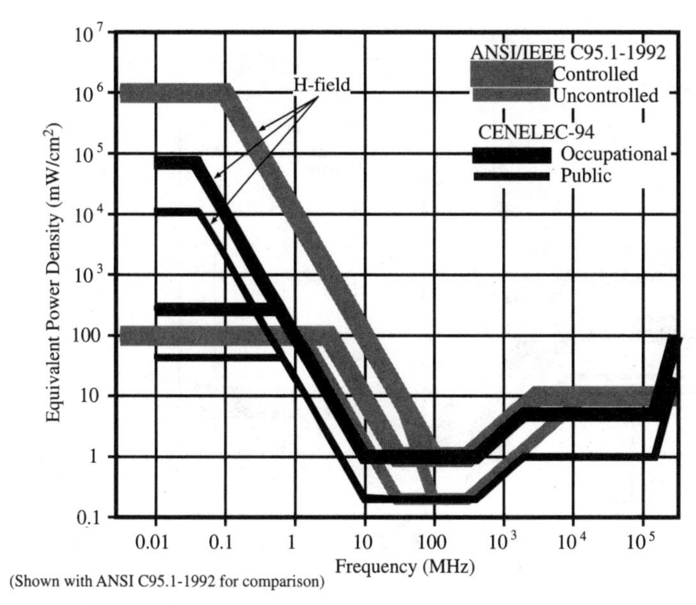

Figure 2.12 CENELEC-94 RF exposure limits.

Ministry's Bureau of Radiation and Medical Devices set forth the first **Safety Code 6,** providing guidelines for limiting human exposure to RF energy. In 1993, the current Code 6 was adopted. Drafted by one individual, Dr. M. A. Stuchly, a well-regarded expert in the field, Code 6 is succinct, leaving to other publications the lengthy review of research literature on which the guidelines are established. The limits of Safety Code 6 are shown graphically in Fig. 2.13 compared with ANSI-92 for reference.

Mexico

Our other continental neighbor, Mexico, has no promulgated standard for limiting human exposure to RF energy, nor is it believed that the environmental and labor side agreements to the North American Free Trade Agreement have been interpreted to apply such limits.

Eastern bloc

Research within the Soviet Union and Warsaw Pact nations has been used to justify Eastern bloc standards that are several orders of magnitude more restrictive than anything for which Western researchers have been able to demonstrate a need. Western researchers have largely been unable to duplicate the findings, especially those alleging

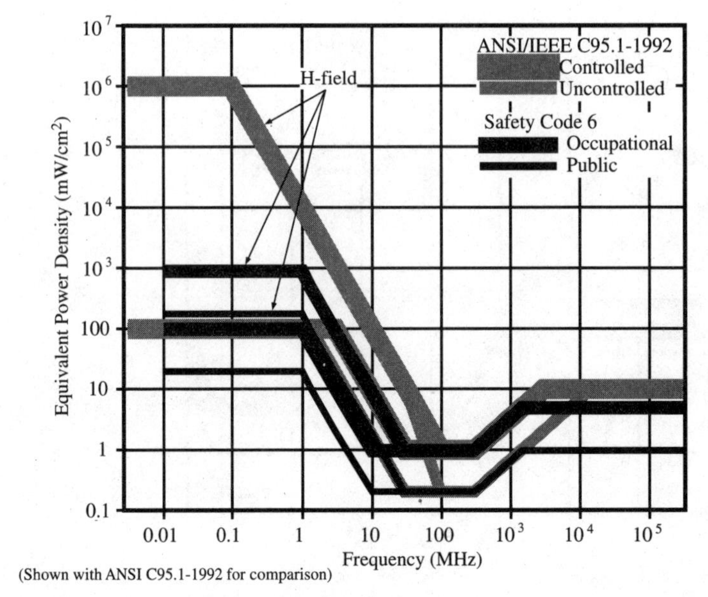

Figure 2.13 Safety Code 6 RF exposure limits.

nonthermal, psychological effects. It is believed that the governments of the Eastern bloc countries made no attempts to comply with their own standards for maximum exposure, some of which[12] are actually *below* ambient levels in many U.S. urban areas. Indeed, the question has been raised as to whether those standards were part of a larger campaign to discredit or confuse the rapid advancements in Western technology being made during that period or perhaps to censor non-governmental broadcasting.

Much publicity was generated in the 1970s about the microwave irradiation of the U.S. Embassy complex in Moscow by its Russian hosts. Such irradiation occurred daily for 22 years, from 1953 to 1976. Until 1975, the highest power density levels were 0.005 mW/cm^2, lasting 9 h per day. For the final 9 months, until February 7, 1976, the maximum power density levels were 0.015 mW/cm^2, lasting 18 h per day and approaching the Embassy building from a different direction. All radiation was in the 2.5 to 4.0 GHz range. Several studies of pos-

[12]The occupational standard in the former Soviet Union has been reported at 0.025 mW/cm^2, with the public exposure level as low as 0.001 mW/cm^2 for many years and 0.0025 mW/cm^2 at the time of the Soviet collapse.

sible health effects on U.S. diplomatic personnel and dependents have been done, the most extensive being the 1978 study by the Johns Hopkins University Department of Epidemiology under contract to the Department of State Office of Medical Services.

That study investigated the health histories of 1,827 persons stationed in Moscow, with more than 3,000 of their dependents, and used as a comparison group 2,561 U.S. diplomatic personnel stationed at eight other Eastern European study posts, along with more than 5,000 of their dependents. Nearly 22,000 individual medical examinations were included, covering 95 percent of Moscow employees, about 80 percent of Department of State employees, and about 40 percent of non-State employees. While cautioning that the study was being done shortly after the irradiation ceased and that therefore latent effects may not yet be evidenced, the study report nevertheless concluded:

> To summarize, with very few exceptions, an exhaustive comparison of the health status of the State and Non-State Department employees who had served in Moscow with those who had served in other Eastern European posts during the same period of time revealed no difference in health status as indicated by their mortality experience and a [variety] of morbidity measures. No convincing evidence was discovered that would directly implicate the exposure to microwave radiation experience by the employees at the Moscow embassy in the causation of any adverse health effects as of the time of this analysis.

U.S. Agencies

ACGIH

The American Conference of Governmental Industrial Hygienists[13] (ACGIH) develops Threshold Limit Values (TLVs) and Biological Exposure Indices (BEIs) on a wide variety of chemical substances and physical agents to which humans might be exposed in the workplace. The ACGIH guidelines are not purported to be legal standards or to apply to nonoccupational settings, as that organization's policy statement, reproduced in Fig. 2.14, makes clear. Nevertheless, the ACGIH limits provide important guidance for municipal, state, and federal health officials. The limits published in 1994 are shown in Fig. 2.15 and compared with ANSI-82 for reference.

---

[13]Note that this organization is not named "...Governmental and Industrial..."

The Threshold Limit Values (TLVs) and Biological Exposure Indices (BEIs) are developed as guidelines to assist in the control of health hazards. These recommendations or guidelines are intended for use in the practice of industrial hygiene, to be interpreted and applied only by a person trained in this discipline. They are not developed for use as legal standards, and the American Conference of Governmental Industrial Hygienists (ACGIH) does not advocate their use as such. However, it is recognized that in certain circumstances individuals or organizations may wish to make use of these recommendations or guidelines as a supplement to their occupational safety and health program. The ACGIH will not oppose their use in this manner, if the use of TLVs and BEIs in these instances will contribute to the overall improvement in worker protection. However, the user must recognize the constraints and limitations subject to their proper use and bear the responsibility for such use.

The Introductions to the TLV/BEI Booklet and the TLV/BEI Documentation provide the philosophical and practical bases for the uses and limitations of the TLVs and BEIs. To extend those uses of the TLVs and BEIs to include other applications, such as use without the judgment of an industrial hygienist, application to a different population, development of new exposure/recovery time models, or new effect endpoints, stretches the reliability and even viability of the database for the TLV or BEI as evidenced by the individual documentations.

It is not appropriate for individuals or organizations to impose on the TLVs or the BEIs their concepts of what the TLVs or BEIs should be or how they should be applied or to transfer regulatory standard requirements to the TLVs or BEIs.

The Policy Statement on the Uses of TLVs/BEIs was approved by the Board of Directors of ACGIH on March 1, 1988.

Figure 2.14 Policy statement on the uses of TLVs and BEIs.

OSHA

In 1971, the Occupational Safety and Health Administration (OSHA) adopted USAS-66, applying this, of course, to limiting occupational exposures to RF. In 1976, that adoption was downgraded to being merely advisory, which status it still has. In 1982, OSHA proposed to revoke the standard (still USAS-66, and not ANSI-74 or ANSI-82) entirely; after deliberating the matter, it decided in 1984 to retain the standard because it may provide useful advice for employers.

EPA

The Environmental Protection Agency (EPA) is an independent federal agency established by the Executive Branch in 1970, with the charge of permitting "coordinated and effective governmental action on behalf of the environment." While this charge does not expressly say the *human* environment, the EPA is looked to for guidance with

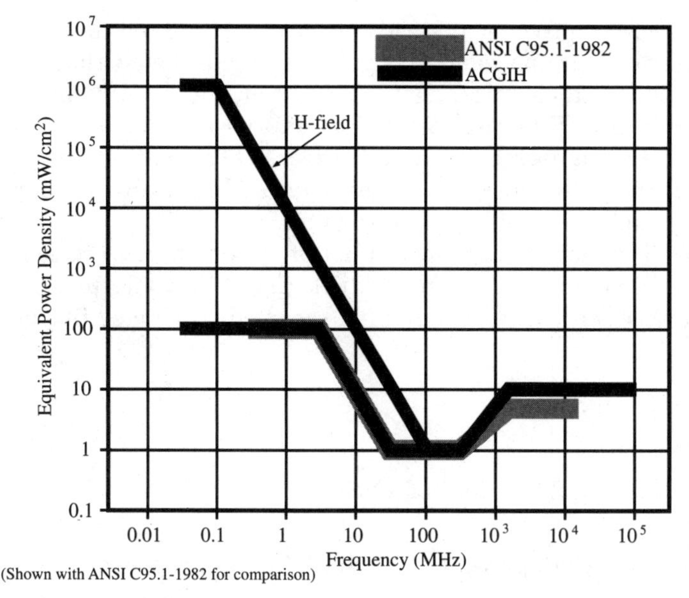

Figure 2.15 ACGIH RF exposure limits.

respect to human exposure to environmental factors. Within its broad jurisdiction, the EPA's activities include water, toxics, pesticides, solid waste, emergency response, air, and radiation, with respect to which the EPA may assume responsibility for (1) research, (2) development of standards, and (3) enforcement of those standards.[14] The EPA's efforts at RF regulation form an involved story of competing offices within the Agency and funding battles outside it.

The question of biological effects of RF radiation has been studied by the EPA's Office of Radiation Programs (ORP) since 1972. By 1979, the ORP was ready to undertake development of "Federal Radiation Protection Guidance" and, indeed, an Advance Notice of Proposed Recommendations was published in the Federal Register in December 1982. The EPA's Office of Air and Radiation (OAR) had concluded that definition of maximum permissible exposure levels was appropriate, and an Interagency Work Group with representation from at least 16 federal agencies was brought together to draft the actual Notice of Proposed Recommendations.

By April 1984, the EPA was reportedly poised to propose a limit that would have been 10 times tighter than ANSI-82 in the body-reso-

[14]Legislator, judge, and jury, as it were.

nance range (*i.e.,* 0.1 mW/cm^2 instead of 1 mW/cm^2), based upon the factor of 10 between occupational and public exposure levels that traditionally applied with chemical agents. However, the EPA's Office of Policy, Planning and Evaluation (OPPE) expressed concerns that ORP had failed to enumerate any actual adverse health effects in humans and that the EPA, since it was not *required* by statute to adopt RF exposure guidelines, should not do so at all.

The OPPE subsequently developed a list of options that were to be floated for public comment and from which the EPA would presumably select its standard. That internal list reportedly included five options:

1. Take no regulatory action
2. Adopt ANSI-82
3. Adopt guidelines at 20 percent of ANSI-82 (similar to NCRP No. 67 and IRPA-82)
4. Adopt guidelines at 10 percent of ANSI-82 (a new proposal)
5. Adopt guidelines at about 2.5 percent of ANSI-82 (similar to Soviet standard)

When finally issued in the Federal Register on July 30, 1986, only the first four options were presented, with the following excerpts from the Federal Register:

> *Option 1.* 0.04 W/kg SAR (a "no effects" level, with uncertain factors of 10 to 100). "This level might be viewed as unnecessarily stringent in that the health protection provided may not be commensurate with its cost, particularly in view of the present uncertainties surrounding low-level and nonthermal effects."
>
> *Option 2.* 0.08 W/kg SAR (also "no effects," with uncertainty factors of 5 to 50). "The Agency is particularly interested in comments on this option, because [Federal law] says that the Administrator shall consult with the NCRP, among others, on radiation matters."
>
> *Option 3.* 0.4 W/kg SAR (equivalent to ANSI-82). Although the "1982 ANSI standard, set at this level, does not differentiate between occupational and public exposure,...[t]he Agency is particularly interested in comments on this option, because it is similar to [that] voluntarily adopted by much of the broadcast industry."
>
> *Option 4.* No regulatory action. "It is not clear whether this approach will meet the perceived need of Federal agencies, States, or industry that have requested uniform Federal exposure guidelines."

Reading only a little between the lines, the apparent leaning of the EPA was for Option 2. The force of established practice meant a strong inclination toward a two-tier standard, while the differential of 5, instead of 10, was justifiable in light of the absence of demonstrable human health effects. IRPA, the State of Massachusetts, and, most importantly, NCRP had reached the same conclusion, and it was expected that the EPA would adopt Option 2. The FCC filed comments in response to the EPA's Notice, urging the EPA to adopt one of the three regulatory options, as well as considering the subsequent establishment of occupational exposure standards.

Incredibly, the EPA closed down its research efforts within ORP on the issue of RF radiation later that year, and the anticipated standard was never issued. Other federal agencies were aghast, as was an expectant broadcast industry. It seemed that the work had been done, the rationale for each option already published, and all that remained was to choose one.

By 1988, the EPA made it quite clear that it would be deferring indefinitely any RF guidance, citing budgetary constraints, lack of specific congressional directive for such guidance, and the press of more urgent issues. To date, the EPA has not regenerated its RF radiation protection programs.

Since the project had already been funded, the EPA published in 1990 an important summary of research to date, entitled "Evaluation of the Potential Carcinogenicity of Electromagnetic Fields (EPA/600/6-90/005B)." While duplicating some of the work done by NCRP in its Report No. 86 (1986), new research published through mid-1989 was included, covering a frequency range from 3 Hz to 30 GHz (*i.e.,* including both ELF and RF). The report's evaluation of the findings from the accumulated studies with respect to alleged carcinogenicity of RF ranged from neutral to strongly negative. A series of meetings was scheduled and held for both internal and external review of the report, but it was not ever issued in final form, with the printed copies still being marked, "Review Draft (Do Not Cite or Quote)."[15]

In 1993, the EPA discovered a renewed expertise and interest in RF exposure standards, as evidenced by the comments the EPA filed in the FCC Docket 93-62 proceeding. Since the EPA had elected not to issue Federal Guidance, the FCC was left to identify and adopt a standard for itself. Selecting the most recently promulgated standard,

---

[15]Of course, everyone proceeded to do both, as the EPA speaks with authority on such matters, and other federal agencies, including the Federal Communications Commission, need to be prepared to follow the EPA's position.

ANSI-92, seemed logical and prudent, and the process of adoption was expected to take only a matter of months. The biggest single factor delaying that process is the objection of the EPA: "EPA recommends against adopting the 1992 ANSI/IEEE standard because it has serious flaws that call into question whether its proposed use is sufficiently protective of public health and safety."

In its comments, filed by the Office of Radiation and Indoor Air, the EPA expressed four complaints about ANSI-92:

1. ANSI-92 relaxes ANSI-82 by a factor of 2 at frequencies above 1.5 GHz.
2. The classifications of "controlled" and "uncontrolled" are "not well defined."
3. ANSI-92 states (incorrectly, in EPA's view) that certain subgroups of the population are not known to be at enhanced risk from RF exposure.
4. ANSI-92 is not fully protective because it is based upon thermal effects.

The EPA made a four-step recommendation to the FCC:

1. Adopt NCRP-86.
2. Add on the induced and contact current provisions of ANSI-92.
3. Add ANSI-92's uncontrolled environment exclusion of devices only below 1.4 W.
4. Ask NCRP to update its standard.

Apparently, the EPA was not bothered by the fact that NCRP-86 and ANSI-92 field exposure guidelines are based on *exactly* the same threshold of tolerance (4 W/kg), that they reference *exactly* the same criterion (interruption of primate behavior), or that they incorporate *exactly* the same safety factors (10 and 50). In many respects, the two standards are identical (refer back to Fig. 2.3*c*). The EPA is objecting, in essence, to the choice of *wording* of ANSI-92, not so much the actual Maximum Permissible Exposure guidelines. In fact, ANSI-92 adopted NCRP's two-tier approach expressly *despite* the consensus finding that "no reliable data exist indicating that...[c]ertain subgroups of the population are more at risk than others." NCRP-86's finding was that "the sensitivity of aged individuals, of pregnant females and their concepti, of young infants, of chronically ill persons is not known." Apparently, the latter wording is adequate, in the judgment of the EPA, but the former is seriously flawed.

Perhaps the EPA was upset that one of its 1986 proposed options was not selected when ANSI adopted the revision of C95.1 in 1992. The irony is that, had the EPA actually done so itself, none of the subsequent confusion, or delays, should have arisen. Probably, few persons have been exposed to high RF fields who might not have been otherwise, but the 6-year delay in promulgation of a standard by some authority other than EPA, and the almost 3-year delay on the FCC action, has meant a 9-year total delay attributable to the EPA's inopportune inaction and action, respectively.

Both NCRP-86 and ANSI-92 recognize that above 300 MHz a transition is made to frequencies that do not penetrate deeply into biological tissue and for which power density may not be a reliable indicator of SAR. A relaxation of power density limits above 1.5 GHz by a factor of 2 is not large in the face of the gross estimations underlying both standards at these frequencies. The EPA presented no data to refute ANSI-92's refinement of that level, based on further research since NCRP-86's 1982 literature cutoff date; in fact, the EPA did not address it, at all, in its detailed comments.

Thus, despite its recognized obligation to set federal policy on matters affecting the human environment, the EPA has not only shirked that obligation but also interfered with the ability of another federal agency to accommodate that failure. The EPA's strong objection to ANSI-92 so delayed the FCC's adoption process that, 2½ years later, it took, literally, an act of Congress (the Telecommunications Act of 1996) to force a resolution.

ANSI/IEEE C95.1-1992

IEEE C95.1-1991

As with previous revisions of C95.1, IEEE Subcommittee IV of Standards Coordinating Committee 28 carried the responsibility for the enormous effort required to develop a consensus standard. The scope of SCC28 in 1991 was:

> Development of standard for the safe use of electromagnetic energy in the range of 0 Hz to 300 GHz relative to the potential hazards of exposure of man, volatile materials, and explosive devices to such energy. It is not intended to include infrared, visible, ultraviolet, or ionizing radiation.

Subcommittee IV's limits of concern were set at 3 kHz to 300 GHz.

Figure 2.16 is a copy of Fig. A7 from IEEE C95.1-1991, showing the flow of responsibility and recommendations. SC-4 had 125 members at the time of balloting on the new Standard, comprising 90 from academic,

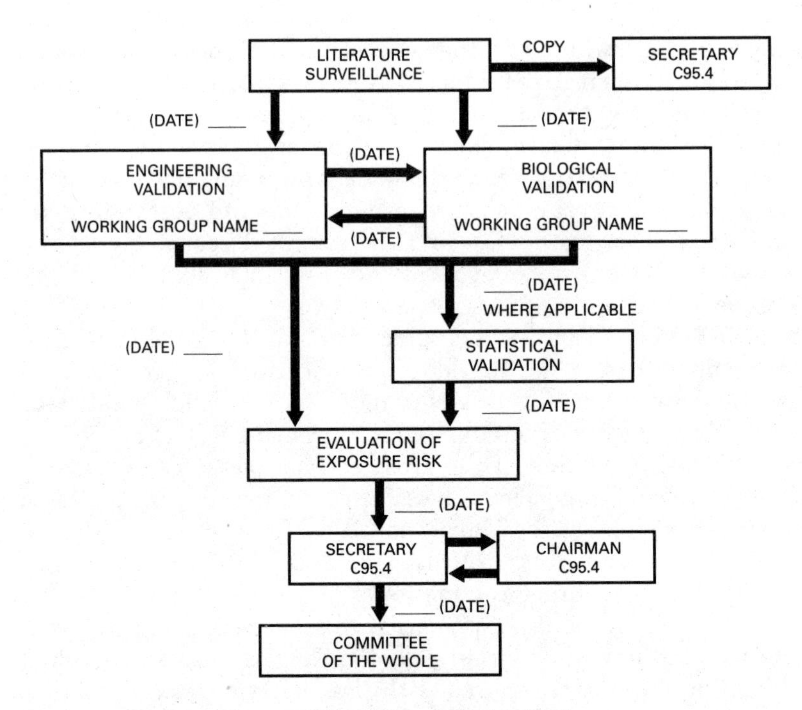

Source: Figure A7 from IEEE Standard C95.1-1991

Figure 2.16 IEEE SCC28 review process.

government, or military research, 16 from industry, 5 from government administration, and 14 independents. The literature search was the most extensive to date and reflects a December 1985 cutoff date for consideration. Both engineering and biological validation had to be made of the study before it was to be included in the database comprising the studies from which evaluation of risk would be made.

Unfortunately, IEEE-91 is not as well crafted as the earlier standards had been. Perhaps SC-4 missed the steady guidance of Dr. A. W. Guy, who had chaired SC-4 for both the 1974 and 1982 revisions of C95.1. The new standard includes various sections that do not necessarily mesh as well as they might, as described below; this is the sign of product put together by a committee of people with disparate concerns, interests, and causes.

Maddeningly, it was an effort to follow the work of SC-4 during its development of the standard. The "final-final" draft was being quoted by outside sources in the trade press,[16] although members of SC-4 were not to release it publicly. IEEE refused for many months even to *sell* a copy of the draft to the author's firm, finally relenting in August 1990. Upon review, it was apparent that several ambiguous or even conflicting provisions had been written into the draft.

Body current limits

The principal concern had to do with the application of the new body current measurements (induced and contact) up to 100 MHz. There were only two papers[17] accepted in the ANSI-92 database that discussed body current measurements. One measured human body impedance, which is necessary for compliance with ANSI-92, only as high as 3 MHz (this is the data that was included in ANSI-92 as Fig. A6 and is reproduced here as Fig. 5.3). The other paper measured induced body currents only as high as 50 MHz. Thus, *there was no scientific basis to justify the extension of body current limits to 100 MHz, nor was there any data by which compliance could be established.* It seems incredible that this provision would have made it into the final draft of the revised Standard, and the author's firm fought immediately, and repeatedly, to have the new body current limits provisions limited to some frequency less than 100 MHz, say, 50 MHz, at least. The concurrently approved IEEE Standard C95.3-1991, *Recommended Practice for the Measurement of Potentially Hazardous Electromagnetic Fields—RF and Microwave,* noted:

> SAR may be assessed by measuring the RF current flowing in an exposed object. In humans, measurement of the induced currents flowing in the legs to ground have been studied at RF frequencies at *below about 50 MHz.* [emphasis added].

Perhaps 3 MHz would have been best, since the necessary impedance data existed only to that frequency. The point is that 100 MHz was simply insupportable.

Even the conditions under which the body current limits apply are ambiguously defined. Induced current limits apply "[f]or free-standing individuals (no contact with metallic objects)," but it is not specified whether this means a *grounded* or *ungrounded* "free-standing individual." Nor is it clear whether induced currents would apply to

---

[16]The first publication of its contents is believed to have been in the 1990 *International Journal of EMC,* issued May 1990.

[17]Both papers were coauthored by one of the SC-4 Co-Chairs, not coincidentally, and accepted after the SC-4 publication cutoff date.

exposure conditions on a tower, for instance; does a tower climber have "contact with metallic objects"? What if the climber is leaning back in a climbing belt to take a rest (keeping in mind the 1-second averaging requirement for body currents)?

Environment classifications

The establishment of new definitions for the widely understood "occupational" and "public" exposure situations has been found unnecessary by many. The use of "controlled" and "uncontrolled" seems to make a distinction without a difference. A variety of definitions are used at different locations in ANSI-92, but generally the *environment* classification is defined in terms of *people*: how much do they know, and what type of control do they have. Also, as discussed in Chap. 3, the "incidental result of passage" clause is so ambiguous, especially with its footnote, "The means for the identification of these areas is at the discretion of the operator of a source," that its definition and even intent are not clear. The result is that, as noted in the Introduction, the industry simply uses "controlled" to mean "occupational" and "uncontrolled" to mean "public."

Measurement techniques

Many of the exposure criteria are actually written in terms of measurement technique, which prejudges, and limits, the manner in which industry would be able to implement the standard. Such recommendations, if they were of concern to individual members of SC-4, should have been included in Standard C95.3 on measurement practice, not in C95.1, which established Safety Levels. Spatial averaging, as an example, was specifically discussed as a measurement criterion in C95.1 §4 Recommendations, which sets forth the exposure guidelines themselves, and in §6 Rationale, which sets forth the scientific basis for the Recommendations. Within §4, relaxation of the power density limits is allowed "for exposure of all parts of the body *except* the eyes and the testes [emphasis in original]." Of course, compliance with this provision effectively negates the spatial averaging provision that was so heavily encouraged (see Chap. 3).

Maximum permissible exposure

The exposure limits of the Standard are shown in Fig. 2.17. Figure 2.3*c* should be reviewed, too; it shows ANSI-92 compared with NCRP-86. The similarity of the two is immediately apparent; the later-issued ANSI-92 does extend the range of the guidelines at each end of

Frequency	Electromagnetic Fields						Body Currents	
Applicable Range (MHz)	Electric Field Strength (V/m)		Magnetic Field Strength (A/m)		Equivalent Far-Field Power Density (mW/cm^2)		Induced (Foot) or Contact (mA)	
0.003 – 0.1	614	*614*	163	*163*			1000f	*450f*
0.1 – 1.34	614	*614*	16.3/f	*16.3/f*			100	*45*
1.34 – 3.0	614	*823.8/f*	16.3/f	*16.3/f*			100	*45*
3.0 – 30	1842/f	*823.8/f*	16.3/f	*16.3/f*			100	*45*
30 – 100	61.4	*27.5*	16.3/f	$158.3/f^{1.668}$			100	*45*
100 – 300	61.4	*27.5*	0.163	*0.0729*	1.0	*0.2*	no limit	
300 – 3,000					f/300	*f/1500*	no limit	
3,000 – 15,000					10	*f/1500*	no limit	
15,000 – 300,000					10	*10*	no limit	

Notes: f is frequency of emission, in MHz.
Uncontrolled Environment limit is shown in *italics*.

Figure 2.17*a* ANSI/IEEE C95.1-1992 Radio Frequency Protection Guide.

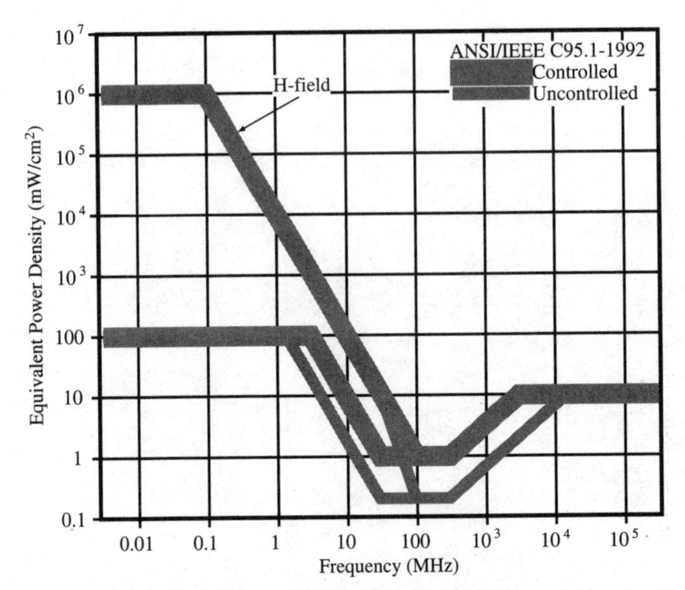

Figure 2.17*b* ANSI/IEEE C95.1-1992 Radio Frequency Protection Guide.

the RF spectrum and also relaxes the limits at those frequencies, recognizing in particular the different manner in which electric and magnetic fields appear to interact with the human body below about 100 MHz (half-wavelength of about 5 feet). Both standards introduce a limit on current flowing in the human body; ANSI-92 included induced and contact current limits while NCRP-86 included just contact current limits (extending to 30 MHz).

ANSI-92 is distinctly not a "thermal standard," a characterization often unfairly attributed to it by otherwise knowledgeable people.[18] There were 14 subgroups of SC-4 making up the Biological Validation Working Group, as follows: 1) behavior, 2) biorhythms, 3) cardiovasculature, 4) central nervous systems, 5) development and teratology, 6) endocrinology, 7) visual systems, 8) genetics, 9) modulation (RF), 10) hematology-immunology, 11) metabolism-thermoregulation, 12) oncology, 13) combined effects, and 14) physiology. The standard was not designed to limit a person's temperature increase to some amount that had been declared unsafe; it is based upon prevention of all reported and substantiated effects, whether they were directly tempera-

[18]This would include the EPA staff submitting comments in Docket 93-62.

ture-related or whether the primary effect was nonthermal. Indeed, the most sensitive measure used to establish the standard was behavioral modification.

Appeal to ANSI BSR

Efforts to modify the C95.1 draft to address these several weaknesses were unsuccessful, and ANSI adopted IEEE C95.1-1991 as its own standard, ANSI/IEEE C95.1-1992, on November 18, 1992. The author's firm then submitted an appeal to the ANSI Board of Standards Review (BSR), and that appeal was joined by C.S.I. Telecommunications, Capital Cities/ABC Inc., CBS Radio, Greater Media, Inc., Group W, Smith & Powstenko, Susquehanna Radio Corp., and the National Association of Broadcasters, representing thousands of stations nationwide. The appeal was heard in New York on February 4, 1993, and submissions were made to demonstrate (1) lack of due process and (2) incorporation of poor science. Surprisingly, or perhaps not, the BSR was not persuaded and 4 days later denied the appeal, leaving broadcasters, in particular, and other users of RF to grapple with how to comply with ambiguous, conflicting, or nonexistent requirements.

Nevertheless, with its adoption by ANSI, ANSI/IEEE C95.1-1992 became the standard that needed to be studied and understood, and those provisions leading to the most conservative result needed to be followed.

Docket 93-62

On March 11, 1993, the FCC adopted a Notice of Proposed Rulemaking (Docket 93-62), by which it would apply ANSI/IEEE C95.1-1992 to all FCC licensees. Extensive comments were filed by numerous parties, reportedly totaling over 3,000 pages (including later reply comments). Specific measurement protocols were proposed (see Chap. 5) to clarify some of the ambiguity in the Standard, and action was expected by 1994, at least.

By 1995, it had become clear that the FCC was under too much political pressure to make any decision at all. The EPA was pressing hard for the explicit adoption of the similar NCRP-86, while the industry was pushing for the more complete ANSI-92. Neither lobbying group had prevailed in the stalemate, but the U.S. Congress forced the FCC to make some decision (*any* decision) by August 6, 1996. Logic had suggested that it would be ANSI-92, but the government's action was a made-in-committee decision to adopt parts of both NCRP-86 and ANSI-92 but neither one completely.

Next revision

As noted above, ANSI-92 is not a consistent standard, and it is a widely held opinion that the next revision of C95.1 should attempt to clean up, at least, the several aspects that caused the most consternation. SC-4 has continued to meet about twice a year, as listed below with highlights from each meeting:

November 12, 1992 (San Diego, California)

Dane E. Ericksen, P.E., Senior Engineer with author's firm, joined SC-4.

June 12/13, 1993 (Los Angeles, California)

Three issues were adopted for balloting:

1. Proposal to adopt induced current exclusion at low ambient field levels
2. Proposal to revise minimum measurement distance
3. Proposal to increase tissue mass for localized averaging

December 1/2, 1993 (Las Vegas, Nevada)

Compromise was adopted for measurements no closer than 20 cm from reradiators and 5 cm from intentional radiators.

Interpretation was given by Co-Chairs that only contact currents are at issue for on-tower work.

Secretary was chided for submittal of comments in FCC Docket 93-62 proceeding without prior circulation within SC-4 for approval.

July 27/28, 1994 (San Francisco, California)

Recommendation was made by Ericksen to relax induced current averaging time to 6 min (since shock/burn was not an issue, by definition); no action was taken.

Previous interpretation was withdrawn regarding induced currents on towers.

November 2/3, 1994 (Baltimore, Maryland)

Issuance of supplement to C95.1 was discussed, with induced current "safe harbor."

June 21/22, 1995 (Boston, Massachusetts)

Balloting was still under way.

November 11/12, 1995 (Palm Springs, California)

Balloting response was still only 55 percent, leading to policy that those members not returning ballots be dropped from SC-4.

June 8/9, 1996 (Victoria, BC)

The Navy reported that 100 mA induced current was exceeded in normal flight deck operations for brief periods, with time-averaged levels 10 to 20 times lower.

R. A. Tell indicated that, above 100 kHz, the issue for induced and contact currents is SAR, and so 6-min averaging would be appropriate. Members in attendance voted unanimously to adopt the longer period.

New balloting was proposed for reaffirming C95.1, revising it, or issuing a supplement.

Summary of ANSI and other standards

By way of summary, the chart in Fig. 2.18 can be helpful to compare ANSI-92 with all the standards previously described in this chapter. Credit should be given, too, for the first introductions of the different aspects making up what is now considered a "complete" standard; Fig. 2.19 lists these regulatory "firsts." Note that ANSI/IEEE C95.1-1992, as the last entry, remains the most complete (complex?) of all standards worldwide. One might be tempted to rhyme,

> Many standards there are,
> With regard to RFR.
>
> But there's been nothing new—
> Not since ANSI-92.

Jurisdictions

NEPA and FCC

By far the largest class of artificial radio frequency emitters, whether intentional or incidental, are those facilities licensed by the FCC. Under the National Environmental Policy Act of 1968 (NEPA), each

Standard	frequency range	underlying SAR limit	"public" 2nd tier	representative guidelines: whole-body resonance	representative guidelines: microwave	representative guidelines: H-field relaxation	whole-body time-average	body current: contact	body current: induced
body-year	*MHz–GHz*	*W/kg*		*mW/cm²*	*mW/cm²*	*low freq.*			
Bell-57	30–100	n/a	n/a	1	1	no	none	no	no
Army-58	30–100	n/a	n/a	10	10	no	none	no	no
U.K.P.O.-60	??	n/a	none	10	10	no	??	no	no
USAS-66	10–100	n/a	none	10	10	no	6-minute	no	no
ANSI-74	10–100	n/a	none	??	??	no	6-minute	no	no
USAF-75	.01–300	??	n/a	10	10	no	??	no	no
Code 6-79	10–300	??	none	??	??	no	1-hr or 1-min	no	no
NATO-82	.01–300	??	n/a	10	10	no	6-minute	no	no
ANSI-82	.3–100	0.4	none	1	5	no	6-minute	no	no
NCRP-86	.3–100	0.4	20%	1	5	no	6-/30-min	yes	no
NATO-88	.01–300	0.4	n/a	1	10	no	6-minute	no	no
IRPA-88	.1–300	0.4	20%	1	5	yes	6-minute	yes	yes
CEC-91	.1–300	0.4	n/a	not specified		no	none	yes	yes
DIN/VDE-91	.03–300	0.4	20%	1	5	yes	6-minute	??	??
ANSI-92	**.003–300**	**0.4**	**20%**	**1**	**10**	**yes**	**variable**	**yes**	**yes**
CEC-92	0–300	0.4	n/a	1	5	yes	6-minute	yes	yes
NRPB-93	0–300	0.4	children	1	10	yes	15-minute	yes	yes
Code 6-93	.01–300	0.4	20%	1	5	yes	6-minute	yes	no
CENELEC-94	.01–300	0.4	20%	1	5	yes	6-minute	yes	yes
ACGIH-94	.03–100	0.4	n/a	1	10	yes	variable	yes	yes

Figure 2.18 Progression of and comparison among various standards.

first public standard........ U.K. Post Office (1960)
first 0.1-hour time-averaging........ USAS C95.1-1966
first partial-body consideration........ ANSI C95.1-1974
first frequency dependence........ USAF AFR 161-42 (1975)
first low-frequency application........ STANAG 2345 (1982)
first explicit SAR threshold........ ANSI C95.1-1982
first two-tier guidelines........ NCRP#86 (1986)
first contact current limit........ NCRP#86 (1986)
first low-frequency H-field relaxation........ IRPA (1988)
first induced current limit........ IRPA (1988)
first frequency-dependent averaging time........ IEEE C95.1-1991

Figure 2.19 RFR regulatory firsts.

federal agency is required to assess whether its actions have an adverse impact on the human environment. The FCC took steps to comply with NEPA by making each licensee responsible for assessing that question and either alerting the FCC to a condition that would have a significant impact or, in virtually all cases, certifying that no such impact occurs. Note that the emphasis is on the *human* environment.

Thus, this certification of no impact can hold even where high fields exist, so long as access to them has been restricted.

The FCC is in a bit of a quandary, however, since it has judged itself not to be an expert agency with regard to health and safety issues:

> [The Federal Communications] Commission is not an expert health and safety agency and must, therefore, rely on expert organizations for guidance on appropriate standards to use.

The EPA's dereliction of its duties in this regard has forced the FCC to look outside the federal government for guidance. Fortunately, the FCC has felt itself to be expert *enough* to issue clarifications for helping broadcasters comply with ANSI-82 and even to take action in Docket 88-469, by which the FCC actually *relaxed* the ANSI-82 requirement for measuring near reradiating objects.

The FCC has continued to claim jurisdiction in this area, with its 3-year-old rule making (in Docket 93-62) to adopt ANSI-92. Once a new standard has been adopted, it is hoped that the FCC will still be an effective forum for obtaining clarifications on interpretation of that standard.

Telecommunications Act of 1996

The Telecommunications Act of 1996[19] contains two provisions affecting the FCC's jurisdiction over RF exposure standards. First, Section 704(b) of the Act requires the FCC to take final action on its proposed adoption of ANSI-92:

> RADIO FREQUENCY EMISSIONS—Within 180 days after the enactment of this Act, the Commission shall complete action in ET Docket 93-62 to prescribe and make effective rules regarding the environmental effects of radio frequency emissions.

The Act was signed by President Clinton on February 8, 1996, 180 days from which was August 6, 1996. While this means that, at last, the FCC has a more current standard, its action accepting neither NCRP-86 nor ANSI-92 in its entirety keeps alive some regulatory uncertainty.

Second, the Telecom Act, as it is colloquially known, preempts state and local governments from limiting the deployment of wireless communications services on the basis of RF radiation. Section 704(a)(7)(B)(iv) states:

---

[19]Title 47, U.S. Code of Federal Regulations.

> No State or local government or instrumentality thereof may regulate the placement, construction, and modification of personal wireless service facilities on the basis of the environmental effects of radio frequency emissions to the extent that such facilities comply with the Commission's regulations concerning such emissions.

Effectively, this should simplify the current rash of local controversies over tower and antenna placement by keeping the focus on aesthetic and land use issues (see Chap. 5).

Local ordinances

Many local jurisdictions became sensitized quite early to the issue of RF radiation (RFR). Sometimes the cause was concern about overhead power lines (operating at ELF frequencies, not RF), sometimes the cause was concern over a particular station that may have been creating interference problems, and sometimes the cause was concern about a proposed new tower. Texas is believed to have the been the first, in 1977, to adopt its own standard, applying only to occupational exposures. Others have followed, including Massachusetts (1983), Connecticut (1984), Arizona (1985), and New Jersey (1987); the counties of Boulder, Colorado (1985), Jefferson, Oregon (1985), and King, Washington (1987); and the cities of New York, New York (1978), Onondaga, New York (1980), Avon, Connecticut (1985), Portland, Oregon (1987), and Seattle, Washington (1989).

Since 1989, when it became clear that no expert government agency would be providing federal guidance, even more local governments have taken it upon themselves to set guidelines for telecommunications systems within their jurisdictions. By now, hundreds of cities and counties, if not thousands, have enacted moratoriums on *any* new development involving wireless communications or have already adopted ordinances restricting such development (see Chap. 5).

Enforcement

Most of the power to enforce RF exposures standards is indirect. ANSI and IEEE are standard-setting bodies. Compliance with their standards is voluntary, with no enforcement provisions whatsoever. They have no structure for enforcement or even monitoring of compliance. The EPA is the only agency that has had mandated to it both the obligation to profess expertise and the ability to take enforcement actions. The EPA has not fulfilled its mandate and has not issued Federal Guidance, which also precludes it from exercising its enforcement responsibility.

FCC

Under the National Environmental Policy Act of 1968, the FCC has an obligation to ensure that its actions do not have a significant impact on the human environment. Presumably, there is some sanction for the FCC if it fails to carry out this obligation, at least implying an enforcement obligation. Since all FCC licensees must renew their licenses periodically,[20] the FCC has used that opportunity to require each licensee to certify that its operations are in compliance with the prevailing standard (PS) that the FCC has adopted for that service. Note that the underlying requirement to comply always exists, so far as the FCC is concerned; it is only certification of that compliance that is not required until the station license is renewed. In addition, the certification requirement is triggered for all applications for new stations and for changes in existing stations.

The Office of Engineering Technology (formerly the Office of Science and Technology) at the FCC has provided helpful information about how broadcasters can check for compliance and bring sites into compliance. FCC Bulletin OST-65 (see Chap. 3) has provided guidelines on checking for compliance, which has both helped broadcasters to comply with the PS and helped the FCC to enforce its regulatory responsibility. That is, FCC staff processing renewal applications can use OST-65 to see if a more detailed analysis of RF conditions is required. There has been some lag between the work being done by consulting engineers and the checking done by the FCC staff. This lag is closing but, for instance, the FCC staff at one time did not check on whether a station certifying renewal was located at an antenna farm, where the additive effect of other stations needed to be considered. This oversight led to some stations at antenna farms, especially those who submitted only limited showings, having their renewals granted while other stations, who perhaps filed a more extensive showing of RFR conditions for the whole site, were required to submit yet more information before the FCC granted their renewal.

The FCC staff now knows to check on whether a station is located near others, and they also know to check for RFR compliance being certified for on-tower conditions, as well. Figure 2.20 is an excerpt from the FCC's new Form 303, for license renewal; it clearly demonstrates that the FCC is checking these issues. In fact, the new Form

[20]TV stations had to renew their licenses every 5 years, and radio stations every 7, until the Telecommunications Act of 1996, which set renewal terms "not to exceed 8 years." In Docket 96-90 the FCC has acted to institute uniform 8-year broadcast station license periods.

CAUTION: Even if you conclude from the use of these worksheets that the RF radiation is consistent with our guidelines, please be aware that each site user must also meet requirements with respect to "on-tower" or other exposure by workers at the site (including RF radiation of one tower caused by facilities on another tower or towers). These requirements include, but are not limited to the reduction or cessation of transmitter power when persons have access to the site, tower, or antenna. Such procedures must be coordinated among all tower users.

See OST Bulletin No. 65 for further details.

Figure 2.20 FCC Form 303 Caution regarding RFR showings.

includes fairly sophisticated decision trees, allowing many licensees to satisfy their renewal RFR requirements simply, without having to retain a consulting engineer except for sites with more than about three stations or with known high fields requiring mitigation (see Chap. 3).

This measure will only be adequate for a short time, however, because of the expectation that the FCC will adopt a revised RF standard that has limits on induced and contact body currents. ANSI-92 sets the limits on body currents, as follows (emphasis added):

> RF current induced in the human body, *as measured* through each foot, should not exceed....

and

> [M]aximum RF current through an impedance equivalent to that of the human body for conditions of grasping contact *as measured* with a contact current meter shall not exceed....

Note that each definition of the maximum permissible exposure explicitly states "as measured," precluding the use of predictive tools, such as knowledge of the ambient RF field, to demonstrate compliance.[21] This is, of course, the rationale behind the introduction of body current limits: the electric and magnetic fields are not adequate for determining the presence of a potentially hazardous situation. Thus, measurements will be required at all AM radio stations, VHF TV stations from Channel 2 through 6, and FM stations below 100 MHz, creating a significant new compliance burden.[22]

When the FCC's action occurs, which is expected for late 1996, the situation of 10 years earlier, when the FCC first adopted ANSI-82, will recur: Consulting engineers, on behalf of individual clients, will devel-

---

[21]IEEE SC-4 has recently voted to adopt maximum electric field limits for which a presumption of compliance with the *induced* current limits may be made. The requirement to measure *contact* currents is not affected.

[22]This represents about 9,800 stations across the United States.

op the means and protocols for complying with the standards, and the FCC staff will modify its checking procedures as the nature of what is submitted begins to shift. The timing is unfortunate, since renewal cycles for both radio and TV have already started, and the FCC may be forced to issue a special request for information from those stations that renewed prior to the adoption of the FCC's new standard.

The FCC also becomes involved in enforcement activities when it receives complaints of RFR problems. For instance, the cooperation of one or two licensees at an antenna farm may not be forthcoming for a RFR compliance program. The FCC's Mass Media Bureau, which oversees radio and TV, would write to all the licensees at the site, requesting a single, comprehensive demonstration of compliance on behalf of all the stations at that site, along with certification by each station that it is complying with the recommended mitigation protocols. The request for this showing is made with a specific threat of license revocation, although it is believed that this threat has not yet been exercised.

The FCC's **Compliance Information Bureau** (**CIB,** formerly the Field Operations Bureau) used to maintain an active presence around the country, enforcing the FCC rules, including the RFR provisions. As recently as early 1996, CIB maintained 38 field offices, each managed by an Engineer-In-Charge and staffed with as many as eight engineers, technicians, and support staff. Since then, however, with the Clinton Administration's policies, the FCC has been actively curtailing its enforcement activities. Only 26 field offices remain, with just part-time staff at six of those. The FCC expects that enforcement requirements can be handled as civil matters through the court system, and it may be that the FCC field presence will be cut even further.

OSHA

The Occupational Safety and Health Review Commission[23] ruled in 1976 that the USAS-66 Standard adopted by OSHA was advisory only and therefore could not be enforced. Part of the rationale had to do with the language used by USAS, involving "should" instead of "shall." Nevertheless, OSHA kept that guideline (10 mW/cm^2, at all frequencies) as an advisory level for employers. Despite that 20-year-old ruling and the lack of anything official more recent than USAS-66, OSHA officials have stated publicly, and for the record, that OSHA inspectors are enforcing ANSI-92 until such time as the EPA adopts a standard of its own.

---

[23]A quasi-judicial agency independent of the Occupational Safety and Health Administration.

Local jurisdictions

Increasingly, one of the most direct enforcement mechanisms is the conditional use permit. The key word is *conditional*; the permit issued by a local jurisdiction is valid only so long as those conditions are being met. When the need to comply with a particular PS or with the FCC's adoption of one is included as one of the conditions of approval, the local jurisdiction is retaining enforcement power over that operation. Should the operation no longer meet that condition (compliance with an RFR standard), the jurisdiction can shut down the operation and, in many cases, sue for damages.

Landlords

Another direct enforcement mechanism is the site lease. Most such agreements have provisions covering safety and hazard abatement, within which RFR could be considered to fall. While it once was unusual for contracts to mention RF exposure or the ANSI standard by name, it is no longer. Inclusion of a specific standard benefits both landlord and tenant, since the landlord knows that prevailing standards are being met, under penalty of lease cancellation and possibly even damages, while the tenant knows that there is a specific "safe harbor" criterion for determining compliance.

For a landlord at a site with one or more towers, where maintenance is the landlord's responsibility, it behooves the landlord to enforce PS compliance for all maintenance activities. While the tower riggers retained for the maintenance work might be content to climb without power reductions that may be part of a mitigation protocol for the site, perhaps set forth in an Occupational Exposure Guide (see Chap. 3), landlords need to enforce compliance via their contracts with the tower riggers, and they need to be prepared to bring in other tower crews, if necessary.

Self-imposed enforcement

Even if the FCC dismisses all of its field personnel, even if OSHA never visits, even if the site is already zoned by local jurisdiction for the use, even if the site lease is based on a handshake, and even if the RF operator handles the maintenance, that operator should *still* enforce compliance by its own staff. Just the potential liability from civil action by employees should be a sufficient incentive for the operator to ensure that all employees with access to the site are familiar with the PS, understand the mitigation protocols that may be in place, and agree to comply with them.

Chapter

3

Implementation of Standards

Measurement Equipment

General-purpose equipment

Any calibrated meter for measuring field strength can be used to assess by measurement a specific set of RF exposure conditions. For instance, the line of field intensity meters manufactured by Potomac Instruments, Inc., includes **narrowband** meters that cover all of the broadcast band. The author has, in fact, performed a complete measurement schedule at several sites using just such meters, recording individually the calibrated signal level of all receivable stations. While this method might be cumbersome, requiring frequent readjustment of the receiving antenna, and while it might be time consuming, requiring the frequent recalibration of the meter, it does work. Careful work should allow reasonable closure (*i.e.*, agreement) to be attained against the summary measurement of a broadband meter.

Also providing *narrowband* capability is a spectrum analyzer, which has the added benefit of making a number of such measurements almost simultaneously and displaying the individual signals at the same time. It is not a true **broadband** device, which would sense all energy across a broad band of frequencies and indicate a single, summary reading, although it provides narrowband information across a broad range of frequencies with considerably more convenience than would a collection of narrowband meters. One also has to account for antenna calibration factors across the band, which will introduce increasing lack of accuracy as the range of frequencies observed at any one time is increased. Nevertheless, using a spectrum analyzer and then recording and summing the individual signal strengths, one can assess RF exposure conditions at one or several locations.

Specialized equipment

A broadband instrument should be used to assess an entire site, or to compare by measurement the effectiveness of alternative mitigation measures. These meters are designed to be equally responsive (within some specified accuracy tolerance) to all frequencies across a substantially broad range of frequencies. One could say that, instead of being tuned to respond to a particular frequency, as antennas normally would be, the antennas in the probe have a sensitivity that is equally poor at all frequencies.

Figure 3.1 lists the several probes currently or recently manufactured by Holaday,[1] Narda,[2] IFI,[3] and EM,[4] the major manufacturers of broadband survey instruments. It should be noted that there are a large number of such probes available for purchase from the manufacturers or for rent from third-party equipment leasing companies, and any one of several may be suitable for use at a site with certain, known RF sources. Figure 3.2 shows the frequency range of these RF probes aligned with an RF spectrum chart from Fig. 1.9 in order to show which probes might be suitable for which type of broadcast setting. For instance, at a site with both FM and UHF TV stations, only certain probes include both frequency bands. Meanwhile, at a site with both FM and satellite uplink facilities, or at a site with both AM and FM stations, more than one probe may be required. In such cases, care needs to be exercised in selecting the proper probes so that too much overlap in range does not occur, which could cause the sum of the readings from the two probes to be needlessly conservative.

Special consideration needs to be made of the need in some cases for separate probes to measure both the magnetic and the electric fields. At frequencies above about 100 MHz, one is normally far enough removed from the antennas that the assumption may properly be made that one is measuring in the far field of the antenna (see Chap. 1). In

---

[1]Holaday Industries, Inc., is located at 14825 Martin Drive, Eden Prairie, Minnesota 55344, 612/934-4920.

[2]Loral Microwave-Narda is located at 435 Moreland Road, Hauppauge, New York 11788, 516/231-1700.

[3]Instruments for Industry, Inc., is located at 731 Union Parkway, Ronkonkoma, New York 11779, 516/467-8400.

[4]Electro-Metrics, a Penril Corporation, is located at 100 Church Street, Amsterdam, New York 12010, 518/843-2600. The EM-4420 Sensing System is designed for OEM use.

Holaday – 4000-series meters			
probe	*field*	*units*	*range (MHz)*
4422	E	digital	0.01 – 1,000
4450	E	digital	80 – 40,000
plus all 3000-series probes listed below			

Holaday – 3000-series meters			
probe	*field*	*units*	*range (MHz)*
CH/HCH	H	A^2/m^2	5 – 300
GRE/STE	E	V^2/m^2	0.5 – 6,000
HSE	E	V/m	0.5 – 1,500
LFH	H	A^2/m^2	0.3 – 10
MSE	E	V^2/m^2	0.5 – 5,000

Narda – 8700-series meters			
probe	*field*	*units*	*range (MHz)*
8721/3	E	variable	300 – 40,000
8722B	E	%IEEE91	0.3 – 40,000
8731/3	H	variable	10 – 300
8732	H	%IEEE91	0.3 – 200
8741	E	variable	0.3 – 40,000
8742	E	%IEEE91	0.3 – 2,700
8752/4	H	variable	0.3 – 10
8760/1/2	E	variable	0.3 – 1,000
8781	E	variable	2,000 – 18,000
8782	E	variable	0.003 – 1,000

Narda – 8600-series meter			
probe	*field*	*units*	*range (MHz)*
8621/3D	E	mW/cm^2	300 – 40,000
8631/3	H	mW/cm^2	10 – 300
8652	H	mW/cm^2	0.3 – 10
8662B	E	mW/cm^2	0.3 – 1,000
8682	E	%ANSI82	0.3 – 1,500

Narda – 8500-series meter			
meter	*field*	*units*	*range (MHz)*
8512	E/H	mW/cm^2	10 – 40
8520	E/H	mW/cm^2	50 – 220

IFI – RHM-series meter			
meter	*field*	*units*	*range (MHz)*
RHM-1/2	E	V/m	0.01 – 220

IFI – EFS-series meter			
probe	*field*	*units*	*range (MHz)*
integral	E	V/m	0.01 – 500
0301/S/H	E	V/m	0.3 – 1,000
3040	E	V/m	300 – 40,000

EM – EM-series probes			
system	*field*	*units*	*range (MHz)*
4420	E	V/m	0.01 – 1,000

Figure 3.1 Broadband field probes.

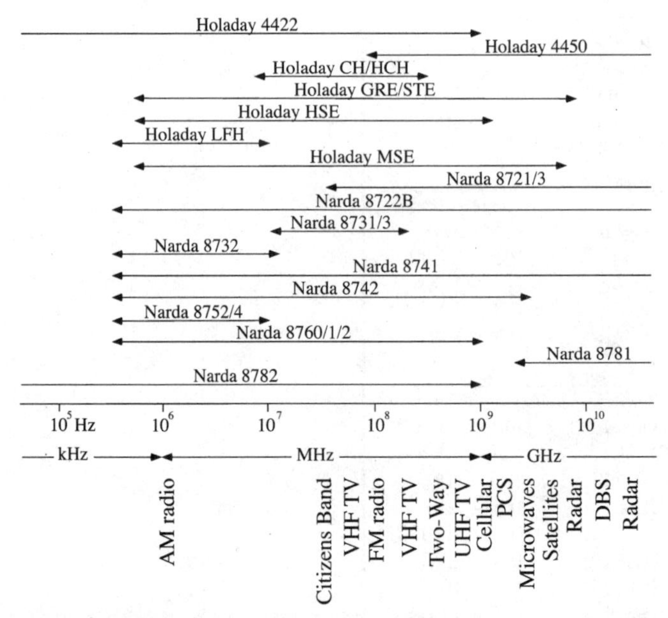

Figure 3.2 Frequency range of broadband probes.

that case, measurement of just the E-field, for instance, is sufficient to determine whether the power density complies with the prevailing standards. At lower frequencies, however, including those used by AM broadcast stations, that assumption is not valid, and separate electric-field and magnetic-field measurements are required, necessitating separate probes. For instance, when evaluating an AM broadcast site, one might use the IFI RHM-1 for electric fields and the Holaday LFH for magnetic fields, or one could use a Holaday 4000-series meter or a Narda 8700-series meter, each with two probes.

Consideration also needs to be made of the sensitivity of the meter-probe combination being selected for a particular measurement task. Most of the meters include several attenuator settings, but prior knowledge of the power density levels expected to be encountered and knowledge of the standards against which the site is to be assessed will ensure that the probes themselves have an adequate and appropriate range. For example, when measuring low RF levels in areas surrounding a typical cellular base station, the highly sensitive Holaday HSE probe might be appropriate, while developing a mitigation program at a broadcast antenna farm would require the use of the less sensitive Holaday STE, which will allow measurement of fields at levels several times the PS. With the 10,000-times relaxation in magnetic-field limits adopted by ANSI-92 for AM frequencies, probe selection is critical.

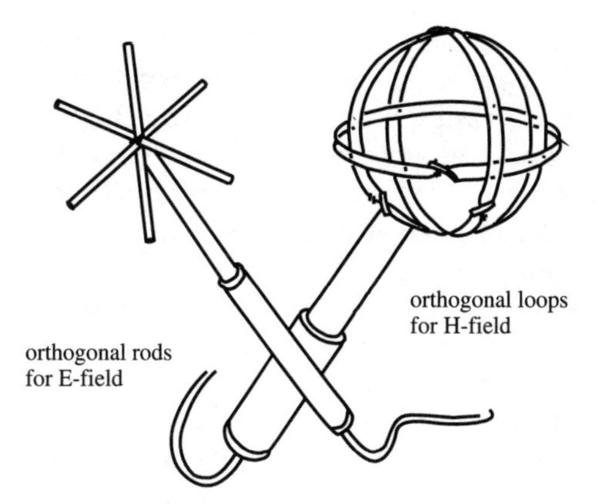

Figure 3.3 Arrangement of E- and H-field probes.

Probe design

Probe elements can be **rods,** which respond to electric fields, or **loops,** which respond to magnetic fields. Figure 3.3 shows the two arrangements. Note that in either case there are three of the elements, oriented orthogonally (at right angles) to each other in three dimensions. This allows the probe to respond to any field in the same manner, regardless of the probe's orientation. **Isotropicity** is the term used for this parameter, that is, the similarity of the probe to an idealized antenna. Broadband probes of high quality will have isotropicity factors of about ±1 dB (about ±25 percent). Some probes have both rod and loops, allowing them to respond to both electric and magnetic fields.

In addition, the detectors within a probe can be one of two different types: diode detector and thermocouple. Diode detector probes have a response that is a combination of linear and "square law" responses. In the presence of two or more signals of approximately equal amplitude, this mixture of linear and square law responses of the diodes will cause the meter to typically read up to 50 percent high, and the theoretical worst-case error could be 100 percent. This makes the error conservative, in that the meter will indicate a greater fraction of the limit (or power density) than is actually present, but it could, without adjustment, cause a site that is actually well under the limit to appear to be exceeding it. If the field strengths from multiple sources differ by 10 to 1 or more, this error becomes small and can generally be ignored.

Thermocouple instruments do not suffer from this problem since thermocouples inherently respond to average power (root mean squared, or RMS). The disadvantage of thermocouple-based instruments is the ease with which the thermocouples in the probe can be burned out.[5] Exposing a thermocouple-based probe to more than about twice its upper power density range may burn out the thermocouples; that is, the inadvertent and momentary positioning of such a probe too near a radiating element of a transmitting antenna is likely to destroy the thermocouples. In contrast, a diode-based probe will generally tolerate fields approximately 800 times the meter's upper limit before the diodes in the probe actually burn out. Thermocouple probes also have had a tendency to drift; that is, they lose their zero reference with changes in ambient temperature. (Further discussion of this problem follows, under "Measurement Procedure.")

Some probes have active electronics in the probe assembly itself. This small "receiver" allows the response of the probe to be exaggerated at certain frequencies and suppressed at others. This ability can be used to shape the probe's response for high sensitivity in the most restrictive range of the frequency-dependent standards; such probes will give readings calibrated in "% of C95.1," for instance, instead of in units of power density or field strength. At sites with RF sources whose ANSI-92 limits are at different levels, use of such a probe can simplify the survey task and avoid overly conservative mitigation measures.

Overall instrument accuracy is normally about ±2 dB, with uncertainties deriving from frequency sensitivity, isotropicity, meter accuracy, and calibration accuracy. For probes covering a wide frequency band, overall accuracy of ±3 dB may be encountered. Accuracy of shaped probes is not necessarily worse than of probes whose frequency response is nominally flat.

Monitoring equipment

The preceding discussion concentrated on calibrated survey equipment, which is needed for many types of measurement programs. There is also a lesser class of RF measuring devices, more properly called monitoring equipment. This type of equipment may have a meter face or digital LED readout but does not have the accuracy of a calibrated survey meter. Sometimes this type of equipment may not even have any readout in the usual sense, using instead an audio alert to signify when some particular power density is encountered. Meters of this type are useful primarily as part of an ongoing occupational mitigation program. For instance, repeated access may be needed to areas of high RF, in which case a personal monitor could be

---

[5]This is particularly true at radar sites, where probe damage can result from the high peak powers despite low average powers.

used to alert such workers when the RF exposure exceeds 50 percent of the PS. The Holaday meter of this type is Model HI-3520, covering 1 to 18 GHz, while the equivalent Narda meter is called a Nardalert, covering 27 to 500, 50 to 1000, or 50 to 2,500 MHz and 1 to 18 or 2 to 44 GHz, depending on the model.

As a second example, periodic checking of fields at a number of installed sites may be a requirement of the permitting jurisdiction or under the terms of a negotiated site lease, in which case a simple meter such as the Alpha Lab[6] TriField meter may be more appropriate. The calibration on this meter is not traceable to any standard, and the manufacturer does not publish calibration data across its response band. Nevertheless, where only gross changes in fields are being sought, for identification of situations needing more detailed investigation with a calibrated survey instrument, the lesser-quality meter may be entirely adequate.

Body current meters

Both of the major manufacturers of calibrated survey instruments (Holaday and Narda) make meters for measuring induced body currents, and Narda also makes an instrument for measuring contact current. The Holaday Model 3701 and the Narda 8850 use parallel-plate sensing, as shown in Fig. 3.4. For obvious reasons, these are known colloquially as "bathroom scale" meters, and their utility for field work is limited.

[6]AlphaLab, Inc., is located at 1280 South 300 West, Salt Lake City, Utah 84101, 801/359-0204.

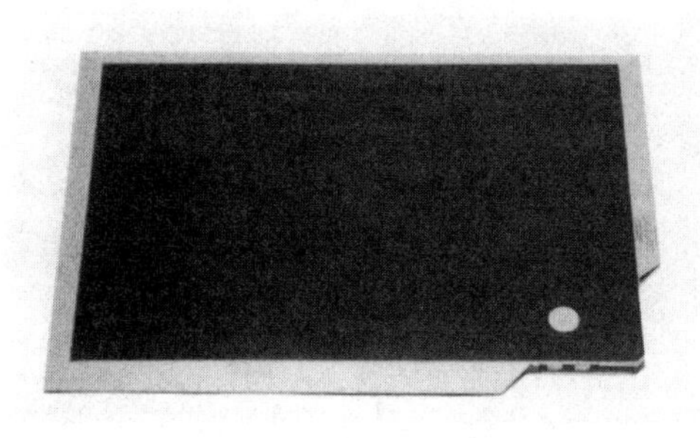

Source: Narda, Holaday

Figure 3.4 Parallel-plate current meters.

For assessing the body current at an operator's position near an RF heat sealer or other industrial or medical device using RF, these meters are just fine, and Narda even makes a version, Model 8870, that is designed as a full-time floor mat, allowing on-going monitoring. However, for areas where people might move around even a little, these meters become too difficult for practical use. The routine would be: (1) bend over and set the meter down, (2) stand up, stand on it, take a reading, (3) step off, bend over, and pick it up, and (4) move to another location, bend over, and begin again. Obviously, this practice cannot be sustained for very long.

As a practical alternative, Holaday has introduced its Model HI-3702 induced current meter, shown in Fig. 3.5. This toroidal anklet meter was first proposed by the author in 1993, and Holaday's product provides a reliable, accurate, and portable means of measuring body current (see Chap. 5).

Equipment for measuring contact current is even more limited. Despite the lack of data on which to base such a meter's response (see Chap. 2), Narda has introduced its Model 8870 contact current meter, shown in Fig. 3.6. Unfortunately, it uses the same configuration as its induced current meter (parallel plates) and so suffers from the same

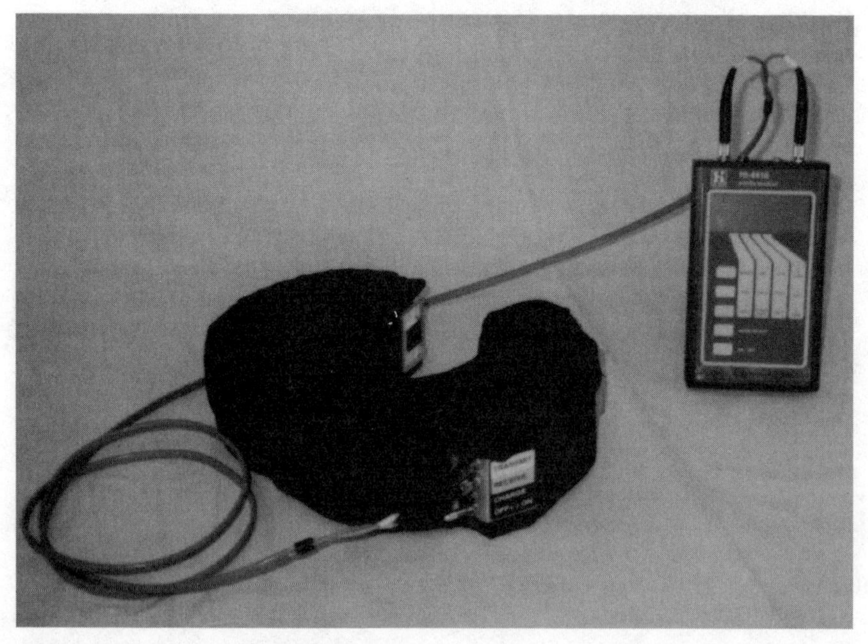

Figure 3.5 Toroidal anklet current meter.

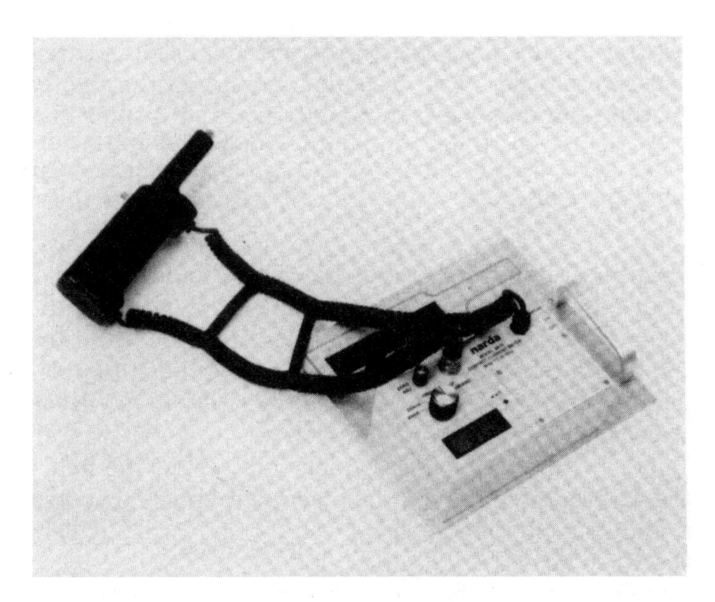

Source: Loral Microwave-Narda

Figure 3.6 Contact current meter.

impracticality for field surveys. (See Chap. 5 for a discussion of new developments in contact current metering.)

Dataloggers

Automatic data recording systems are good at accumulating a lot of data with little operator effort. Commercially available dataloggers record field strength (or power density) as a function of time, and the new Narda Model 8718 digital meter will record minimum, maximum, and average readings over some specific period of time. However, the actual utility of dataloggers depends on subsequent operator effort to extract the data and draw useful conclusions. For RF radiation (RFR) site surveys, the quality of the data recorded is more important than the quantity. For instance, knowing that the ANSI limit was exceeded by X percent at time Y of the survey is hardly adequate, because the ultimate question is, "How is that high field to be mitigated?" Knowing that it occurred at some point in time does not help at all to answer that question; it does not even indicate *where* the high field is.

Even if it were also known somehow that the reading had been taken at location Z, the ultimate question still cannot be answered. Without knowing how big the affected area is, and where it is located relative to possible human access, the only essential issue, mitigation, cannot be addressed. Perhaps if the *entire* area were mapped, in three dimensions and at a resolution in all axes approximating the size of the probe, and if the computer were to reconstruct and display the fields, then.... The critical questions can be answered only with real-time probing of high fields, allowing real-time assessment of the extent of such fields and real-time decisions as to mitigation measures.

Dataloggers may indeed have a use as part of a monitoring program put in place after a proper RF survey has been completed and a mitigation program has been developed. In public areas near an antenna farm under a different jurisdiction, for instance, a technician could be trained to log data along a certain path at regular intervals, and that data could be compared with the original study to check for unanticipated increases in RF power density levels. Or a worker could carry the meter with a datalogger attached to review the fields to which exposure was required during some period of time; far better for this purpose, and with nothing to carry, would be a Nardalert personal monitor, described above.

As discussed more fully later, the use of dataloggers for spatial averaging is a poor engineering practice. Knowing the average power density within some human-sized volume is of no direct benefit; the maximum field within that volume could well have exceeded the PS, and the averaging feature of the measurement equipment conceals that possibility. If such a high field exists, mitigation of that field must be undertaken, unless it can be shown that 1) localized power density limits are not exceeded *and* 2) neither eyes nor testes can be positioned in that area. If spatial averaging is done, as facilitated by a datalogger, neither piece of essential information is provided. There is simply no substitute for a proper field survey by a professional.

Measurement Procedure

Establish the environment

Selection of the appropriate procedures for a particular measurement program will depend on the goals of the program (*i.e.,* what is the question for which an answer is being sought?). For instance, the question might be simply, "Does the site comply with ANSI?"[7] This particular question is among the easiest to answer, since it implies

[7]Since ANSI/IEEE Standard C95.1-1992 is the most recently adopted U.S. exposure guideline, its specific terms are used in sections of this chapter as the representative standard for evaluation purposes. In its place could certainly be substituted NCRP or any other standard, with appropriate relaxations for any requirements that may not be part of that older standard.

that the evaluation standard has been accepted and, especially for an affirmative answer, only minimal data need be recorded. With experience, one learns to recognize which areas of a site are likely to have the highest fields, and so a survey of the type needed to accomplish this limited goal can concentrate on those areas.

While the exposure standards are based upon limiting the specific absorption rate of RF energy, the specific absorption rate (SAR) is not a directly measured quantity, and power density (or field strength) limits are specified for field assessment. Therefore, for any measurement program, the issue is not specifically, "Is it safe?" but more properly, "Does it comply with ANSI?" Given the safety factors expressly incorporated into ANSI, presumably an affirmative answer to the latter question implies an affirmative answer to the former, as well, but the distinction is worth understanding.

The goal of a measurement program at any site should be conservative: to find the highest fields. Since a mitigation program will probably be developed based upon the measurements, it is important that no other person be able to visit the site and find, unless conditions have changed, a higher field, for which adequate mitigation may not have been developed.

One of the first steps in assessing RF exposure conditions at a site is to determine whether the area is classified as controlled or uncontrolled. Recalling the definitions from Chap. 2, these are virtually the same as occupational and public. If people who have access to an area are not aware of the RF levels or cannot control their exposure, the site is uncontrolled. It is important to make this determination before measurements begin so that the appropriate threshold levels of exposure are used for evaluating any high fields that may be observed.

It is also important to have calculated beforehand the relative contributions from any auxiliary antennas that may be installed at the site (even if unlicensed). This is because one may contribute more than the station's main antenna to the power density levels in certain areas of the site. If an auxiliary antenna, or several in combination, needs to be activated to establish worst-case conditions, appropriate arrangements need to be in place before the visit to the site is made. Such measurements should be *in addition to* the measurements of normal operating conditions.

Appropriate meters for the task should also be selected beforehand. As discussed in the preceding chapter, more than one meter-probe combination is normally required to cover both electric and magnetic fields and, perhaps, to cover the range of power density levels expected. Both diode and thermocouple detectors have the potential for out-of-band responses, and awareness of the manufacturer's instructions and cautions will help avoid situations of erroneously high readings. The meter reading corresponding to the ANSI limit (if not a shaped probe) should be known beforehand, of course, for the frequencies of interest, so that an immediate determination can be made as to whether a given field does or does not exceed the limit.

As noted in the preceding chapter, meters may give indications of RF strength in several possible units: percentage of Standard, mW/cm^2 (power density), V^2/m^2 or A^2/m^2 (field strength squared), and V/m or A/m (field strength). In order to determine the meter indication corresponding to the PS limit, conversions may be necessary. A constant relationship between the electric and magnetic field components is assumed, given by the impedance of free space, 377 ohms. In the formulas below, the electric field strength E is in V/m, the magnetic field strength H is in A/m, and the equivalent far-field power density S is in mW/cm^2:

$$S = \frac{E^2}{3770} = 0.000265 \times E^2 \qquad S = 37.7 \times H^2$$

$$E = \sqrt{3770\,S} = 61.4\sqrt{S} \qquad H = \sqrt{\frac{S}{37.7}} = 0.163\sqrt{S}$$

While not normally an issue each time measurements are conducted, a conscious decision should be made as to what "limit" is actually used, in light of the measurement uncertainty due to meter inaccuracies. Since typical meter accuracy (see Chap. 1) is about ± 2 dB, perhaps the "limit" against which observed fields are being evaluated should be the ANSI limit *minus* the meter tolerance. For instance, 0.20 mW/cm^2 (27 V/m) minus 2 dB equals 0.13 mW/cm^2 (21 V/m). However, the measurement procedures recommended in this chapter are generally sufficiently conservative that these lesser values need not be used.

Ambient fields

Except in rare circumstances, the highest fields at a site will be near active antennas, transmitters, and tuning circuits. Therefore, all areas below and in front of antennas should be examined with extra care for ambient fields exceeding ANSI. The probe of the calibrated survey instrument should be moved slowly from ground level to about 6½ to 7 ft above ground, while watching the meter indication throughout. The hallmark of good measurements is **repeatability,** and the highest readings observed (*i.e.,* those most likely to be reported and relied upon for analysis) should be tested several times. The probe should be moved into the high field and then out several times, checking for repeatable readings.[8]

Conservative engineering practice dictates that the space around each high field be tested for the presence of any higher fields. Reported fields should be the highest *observable* at the site so that no later observer, perhaps someone not as well versed in taking meaningful readings, could find any that measure higher and thus cast doubt on the original measurements. All areas exceeding ANSI, and the extent by which they exceed ANSI, should be identified and recorded so that appropriate mitigation measures can be developed, as discussed in more detail below.

Given that one must generally be close to transmitting antennas in order to exceed the occupational limits, typical locations for a survey of this type are mountaintops, rooftops, and AM stations. Considering localized hot spots later, the presence of a high field implies the existence of a location at which ANSI is just equaled. For areas of high fields larger than about 3 ft, the goal of the measurements will be to map the **contour** of all such points. Outside that contour, ANSI is met and no special restrictions are required, while inside it, ANSI is exceeded and some mitigation measures must be implemented in order to ensure that exposures exceeding ANSI do not occur.

The ANSI-limit contour should be carefully identified by moving the probe horizontally back and forth across the ANSI limit and dropping markers on the ground. Surveying or landscaping flags work well for this purpose. Permanent markings should not be made until the location of the entire contour has been established, and the process of finding the ANSI limit and dropping flags should continue until that has been accomplished.

At that stage, the marked location of the contour should be checked for reasonableness. Considerations should include shape (expected to be relatively smooth), position (likely to be under or otherwise associated with one or several antennas), and symmetry (common, although certain antennas are known to have nonsymmetrical patterns). Further measurements and adjustments may be appropriate if some of these characteristics are not reasonable. The area outside the contour should also be checked to be certain that high fields do not recur beyond that line. Then, and only then, should the contour be recorded permanently. Bright surveyor's paint is ideal for this purpose, although such markings might not be possible at every site. In those cases, at least, it is recommended that the contour also be defined with

---

[8]It is noted that dataloggers cannot be used for this purpose, and spatial or time averaging does not apply.

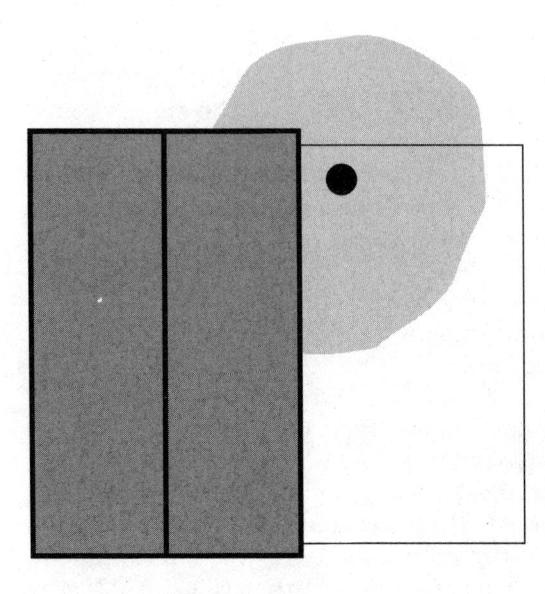

Figure 3.7 Example of high ambient field outdoors.

reference to fixed structures at the site. This way, even if the markings should fade, the contour can still be relocated. Figure 3.7 shows a map of high fields as might be drawn for a site. Note the definition of the contour location around the base of the pole. This data will be used in developing an appropriate mitigation program, discussed later.

Next would be a determination of how high the fields are inside the identified contour. Similar assessments at various locations within the contour may be warranted, depending on the field strengths and what type of mitigation program is being considered. For instance, parts of the area in question might be considered areas of transient passage or areas to which access can be restricted easily. It is essential to have a suitable mitigation scheme in mind before concluding the measurements.

Localized fields

The next step is to check for potential **hot spots.** These are areas of localized high fields, generally associated with **reradiating structures,** whose existence may or may not imply that ANSI is not met, but whose character needs to be well defined before that determination can be made. Generally, a grounded metal object at least several feet tall will cause some reradiation; objects much shorter than that may be too inefficient as reradiators to be of concern. Each potential reradiator should be examined with the probe for the presence of any fields approaching ANSI. Figure 3.8 is a photograph of a publicly ac-

Localized fields exceeding ANSI C95.1-1992 were observed at locations on metal railing.

Figure 3.8 High localized fields.

cessible lookout near some high-power RF sources. At several locations along the handrail, hot spots were observed with dimensions exceeding 20 cm.

If any high fields are found, the **measurement distance** from the object becomes an important consideration. Unfortunately, there has been conflicting guidance on what that measurement distance should be, and its selection may depend upon the particular circumstances. ANSI-82 had specified that measurements should be taken no closer than 5 cm from an object, which experience has shown would lead to the identification of many localized fields that exceeded the ANSI field limits but that did not have the ability to cause violation of ANSI, due principally to their small size and to the likelihood of detuning the object by touching it and thus destroying the field. In Docket 88-469, the FCC provided guidance to the broadcast industry for achieving compliance with ANSI-82 (its then-current standard) and recommended that measurements be made no closer than "10-20 cm" (implying 10 cm, since no guidance was provided for distinguishing what might justify the larger distance). Then, ANSI-92 adopted 20 cm as the recommended minimum distance, and this has been found to remove from consideration virtually all of the hot spots that had made strict compliance with ANSI-82 so cumbersome at many broadcast sites.

It is noted that NCRP echoes ANSI-82's 5-cm distance in this regard, while the CENELEC Prestandard actually specifies minimum measurement distances by frequency:

30 cm for 10 to 100 kHz

25 cm for 100 kHz to 3 MHz

15 cm for 3 to 10 MHz

10 cm for frequencies over 10 MHz

Any hot spots that are observed should be checked for size. If pulling the probe back from the reradiating object causes the field to drop below ANSI at a distance not exceeding the minimum distance, the hot spot need not be considered further. If it still exceeds ANSI at that distance, however, its presence must be recorded. Both the magnitude of the field at the minimum distance and the distance at which the ANSI threshold is reached should be recorded, along with the location, in order to develop appropriate mitigation measures.

Body current

As noted in the previous section, commercially available equipment that can measure induced and contact body currents is limited. Therefore, measurement procedures are only now being established. The exception to this situation is for operators of RF sealing and welding equipment, who are exposed to RF energy while relatively stationary. The bathroom scale meters work well for this situation: stand on it, turn on the RF, and record the reading. All possible operator positions should be examined, since the high fields can vary considerably with changes in position of only a few feet. It is preferred that the regular operators be the "source" of the current in the measurements, since typical operator positions may not be obvious otherwise. At each measurement location, several cycles of the machinery should be run, for each common operator position, such as holding material, pushing the control buttons, or checking the temperature, depending on the process. The highest readings should be recorded for each measurement location. If the operators use the machinery with shields or other protective devices removed or disabled, care should be taken to check *each* reading for currents that exceed the standard so that the tests may be immediately discontinued if they do.

For situations at broadcast sites, assessment of body currents becomes more difficult. A parallel-plate "boot" or the Holaday HI-3702 toroidal anklet meter are the only practical alternatives for induced current measurements. The meter should be attached to the engineer's body and the site should be checked, for the third time,[9] for high body current readings. Proper measurement protocols must be used (see Chap. 5); otherwise, the readings may not be the maximum that could be observed, and repeatability will be poor. Highest induced current readings are expected to be in areas of highest ambient fields, so other areas of the site can be checked relatively quickly. The foot whose current is being monitored should be placed in all high field areas at spacings of about 5 ft, with the engineer assuming the standard test position at each measurement location. Areas near the location(s) with the highest readings should be examined further, until the highest possible reading is obtained. Only in limited situations will this value exceed the applicable limit, but it must be noted that the averaging time for body currents is only 1 second.[10] Thus, mitigation measures will likely be different, and body currents must be evaluated even within areas with high fields whose mitigation procedures involve short durations.[11]

For assessing compliance with the contact current provisions of ANSI, safety of the engineer taking the measurements is a special concern, because high voltages are being sought. Measurement procedures will be heavily dependent on the specific apparatus used. The one commercially available meter, the Narda Model 8870 shown in Fig. 3.6, is straightforward to use: stand on the plate, hold the "gun" against the potentially conducting object, and pull the trigger. (Contact current measurement procedure is discussed further in Chap. 5.)

Narrowband measurements

There are a number of reasons why the measurement program might require narrowband measurements. For instance, if it is believed that high fields in an area are being caused by one or several RF sources out of many such sources at the site, narrowband measurements can be used to confirm that fact and to identify the offending sources. Or, if there are widely different frequencies in use at the site, the basic question of compliance with frequency-dependent exposure standards may hinge on knowing the relative contributions of the individual sources.

---

[9]Ambient fields are checked first, followed by localized fields.

[10]This is a useful factor only at radar sites; at other locations, 1 second is essentially instantaneous.

[11]Of course, the engineer making the measurements must follow those mitigation measures, too. That is why the order of measurements (power density, then body current) is important and why it is important to develop proposed mitigation protocols as part of the measurement procedure itself.

As discussed in the section on measurement equipment, a variety of narrowband equipment might be used. Narrowband measurements consume considerable time if they are to be done properly, so it is a good idea to select one or just a few locations at which they will be made. Depending on the need, a location may be selected near an identified contour nominally exceeding the prevailing standard (PS), so that the relative contributions of different sources can be determined. It may be, however, that the location is selected for simple convenience, noting such factors as wind, sun, and proximity to power.

The procedure using narrowband meters is simply to take a measurement at each frequency in use at the site, recalibrating the antenna each time and moving the antenna about so as to find the maximum field. Each polarization (horizontal and vertical) should be measured and recorded separately. For measurements with a spectrum analyzer, a square should be marked on the ground first, with a side no less than one-half wavelength of the lowest frequency expected to be encountered[12]; the entire volume of a cube over that square is to be swept, in three dimensions, for each frequency range to be checked. The analyzer sweep should be adjusted so that the entire band can be measured with about four or five passes. The analyzer should be kept in peak hold mode for each sweep of the volume, which is also done once for each polarization. The discrete RF signals can be read from the screen (or a printout) and recorded, with proper adjustment for the calibration factor of the antenna used. It is important to readjust the antenna for each frequency range so as not to introduce additional measurement error.

With either procedure, the highest reading for each frequency (range) should be summed to give the total RF field strength at that location. This will produce a conservative result, in that the highest readings probably did not coincide within the volume examined, but there is no other practical method for developing a broadband figure with narrowband meters.

Inside transmitter buildings

Finally, measurements should be taken inside each of the buildings or other segregated areas in which the transmitters or antenna tuning units are located. These spaces are almost always controlled environments, and the ambient fields, localized fields, and body currents can be mapped just as the outside areas were, with contours marked and

---

[12]This will be too difficult, obviously, when a frequency below, say, 88 MHz or 54 MHz is involved, and other methods must be used.

highest readings noted for development of a suitable mitigation scheme. It is important to check, especially inside transmitter buildings, for unterminated lengths of transmission line; these might be connected to an out-of-service transmitting antenna that has become an efficient *receiving* antenna, transmitting the energy directly into the building and radiating it from the open end of the line.

Time and space averaging

Although ANSI contains time- and space-averaging provisions, they should not be used to modify the measurement program for a site; they should be used instead, if at all, in the development of appropriate mitigation measures. To use the time-averaging provision, as an example, unless one can be assured that transit through an area of high fields must be brief (an elevator, perhaps), one cannot rely on such fortuitous passage. (And, even then, a mitigation program for the site might include installation of circuitry to take the transmitters off-line if the elevator were to be halted at that location for some length of time.)

ANSI-92 gives a generous 30 min for time-averaging uncontrolled exposures. However, the essence of *uncontrolled* is that persons unaware of the presence of the field would remain in its presence for a period of time that the operator of the RF source does not control. Likewise, with the space-averaging provisions, ANSI takes away the potential benefit by excluding eyes and testes; one cannot be assured that, for instance, a male hiker would not rest against a fence post with a high field about 2½ ft above ground (this is typical at sites with FM stations). There is, however, no need or benefit for the survey measurements themselves to be averaged over time or space.

General guidelines

Applicable to the measurement procedures just described, as well as to all professionally done measurement programs, are a number of procedural guidelines that should be followed to the extent possible:

1. *Research the site and local RF sources beforehand.* Sometimes areas of high fields are not contiguous or even adjacent; this will ensure that such areas are not missed. It is also important to follow exposure guidelines for oneself (*i.e.*, knowledge of expected fields will help avoid exposing oneself to high fields before they can be fully assessed). Finally, certain meters may be susceptible to damage in fields exceeding their rated capacity; loss of one's meter in the middle of a measurement program would be embarrassing, at the least, and would obviate those (and any subsequent) mea-

surements, at worst, if the failure of, say, only certain sensing elements were not immediately apparent and the meter continued to be used for additional measurements.[13]

2. *Confirm before taking measurements that RF sources are actually in operation and that their operating powers are as authorized.* Broadcast stations may run at something other than their licensed power, and two-way systems (including cellular telephone operations) normally operate at only partial capacity. Without knowledge of what specific conditions exist during the measurements, application of the data to worst-case conditions (*i.e.*, full-power operation) is difficult. If the operating levels were actually to vary in some unknown way *during* the measurements, use of the data would be virtually impossible.
3. *Be careful where the probe goes.* Since most sites involve high-power transmitters or electrical equipment, it is important for one's own protection that concentration not be devoted solely to the meter face during measurements. A habit should be developed of observing the probe location at full reach first, before monitoring the meter. In addition, since access may be provided to the operating rooms of operating stations, one does not want to hit buttons or knobs inadvertently with the probe, running the risk of taking the station off-air or otherwise adversely affecting its operation.

It is good engineering practice to plan the measurement program beforehand in sufficient detail so that surprises (which will happen) do not catch the engineer unprepared. To that end, a checklist similar to that in Fig. 3.9 can be helpful. Of the "items to take" not already discussed, the batteries are to replace the ones in the meter, since there is little more embarrassing that can happen in field work than to have the meter fail due to worn-out batteries. The business cards can save considerable time, should the measurements attract attention and on-lookers begin asking questions. Answering a few questions is good public relations, but a few can easily become too many. A polite, "I have to complete this job and then move on; here's my card...." can be effective without being rude. Water, of course, is always good to have, for personal comfort. Finally, in case issues arise needing specific citations, a copy of the PS may be helpful.

The final two steps are often overlooked, which is why they should be on the checklist. Before departing the site, it is vital for two reasons to have a viable mitigation program worked out for each field or

[13]It is for this reason that Narda's thermocouple probes, which are most susceptible to burnout in high fields, have an autotest feature.

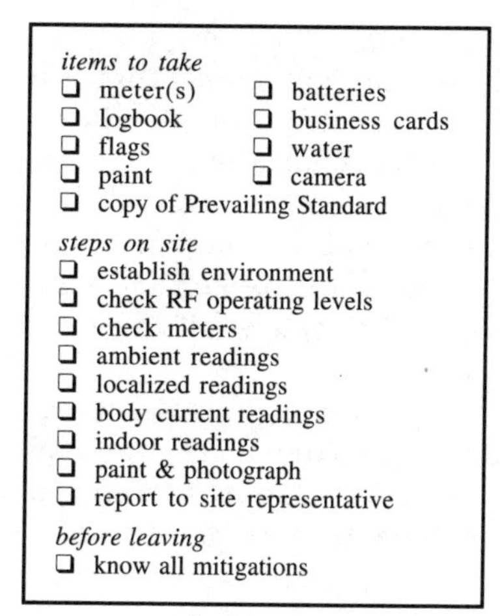

Figure 3.9 Checklist for RF survey measurements.

current requiring it. First, if additional information is needed, it can easily be acquired, but if that need is discovered later, its acquisition may be difficult or expensive. Second, when the task is completed, a verbal report needs to be made to the person representing the site management (or the client, if it is a different entity). That report needs to include the mitigation schemes for all areas needing them so that they can be implemented immediately. Waiting 2 weeks for the report (or even 2 days) may be too long if work is waiting to commence or continue at the site.

Residential measurements

Satisfying concerned neighbors is perhaps the most challenging type of measurement program, but it can also be the most gratifying, in the pleasure derived from educating people on a technical subject, from seeing them come to understand something new, and from witnessing their relief at accepting the (generally) reassuring results. The question being answered is often stated as, "Is it safe here?" However, unless one is willing[14] to answer that bare question, it should be restated as, "Are the standards any good and does this area comply with them?"

The measurement procedures to be followed in this situation are

[14]An engineer would probably not claim to have verified independently the findings of ANSI, NCRP, or one of other standards-setting agencies. Rather, the engineer would profess an understanding of those standards and would claim proficiency both at taking appropriate measurements and at evaluating those results against the standards. A medical doctor, on the other hand, might wish to claim expertise in the evaluation of the basic research and assert, on his or her own authority, the validity of the standards.

much like those described above for transmitting sites. Here, it is certainly the most restrictive, uncontrolled exposure guidelines that are being tested. Calibrated broadband survey instrument(s) should be used, preferably models with an analog (needle) indicator, since this seems to provide a more intuitive indication of the results than does a digital readout. The first step should be, with the concerned neighbor present, assembly of the equipment, followed by a statement of what the applicable limit is (in the same units as the meter will indicate) and an explanation of how the meter works.

After the neighbor gives some sign of acceptance or understanding, the next step recommended is to hand the instrument (meter and probe) to the neighbor. This is normally unexpected, but it is almost essential for the neighbor to develop an appreciation for the measurement procedure. Without that step, derivation of the results remains a mystery, much like a physician saying, "Hmmm..." and then simply pronouncing a diagnosis or prescription. By letting the neighbor make the measurements, there is no question of selective examination to achieve some desired result, and the neighbor may take measurements in unusual places (an upstairs reading loft, for example, or a baby's napping area) that might not have been checked otherwise, but without which the neighbor would never really be satisfied.

Even in group settings, letting a neighbor handle the meter helps remove the mystery. Someone normally comes forward as the spokesperson upon one's arrival; that person is the logical one to take the meter first, and then others in the group may be less reticent to ask for the meter from that person. While the meter is going around, even from person to person, the engineer should stay involved by asking the neighbor what reading is being obtained, by recording those readings in a log book, and by suggesting likely places for additional readings to be taken (the survey would be incomplete without having measured those, anyway).

Of course, certain neighbors are difficult to interest in what the measurements might show, and even careful steps will not elicit an acceptance of the procedure or the results. In that case, one can only verbalize the process being followed and offer them, periodically, the instrument to hold or, at least, the opportunity to suggest measurement locations.

When the survey is complete, or the neighbor is ready to relinquish the instrument, one should verbalize the results (*e.g.*, "The highest reading you observed is *X*, which is about *Y* times below the Standard, which, you'll remember my saying when we began, is *Z*").

The other important step, once the survey of ambient RF levels has been completed, is to allow the neighbor to examine other sources of

RF that may already be in the home. This includes microwave ovens (measure around the door seals and in front of the window), wireless nursery monitors (measure at the transmitting end, near the crib), cordless telephones, and cellular/PCS handsets. This also provides reassurance that the meter is working, should the ambient fields all have been too weak to give a meter indication. Generally, amateur radio operators will not request a visit of the type just described but, if they should, one would also want to examine the fields created when their equipment is transmitting.

Measurement of household appliances (*e.g.,* electric blankets, hair dryers, electric stoves, space heaters) should *not* be done, since the 60 Hz power being radiated is almost certainly beyond the calibration range of the instrument. The distinction should have been made, too, that 60 Hz extremely low frequency (ELF) is not the same as RF, and this provides an opportunity to reinforce that concept.

Calculation of Fields

Simple cases

At one time, the tables and figures published by the FCC in its Office of Science and Technology Bulletin No. 65, issued October 1985, provided useful information for doing simple "calculations." These consisted of scaling a single station's power and looking up distances in tables or figures. However, that information was based on ANSI-82 and so does not reflect the public (uncontrolled) nature of later standards. As a consequence, their application should be limited now to occupational settings; few are left that are truly so simple. It is expected that the FCC will provide similar guidance following adoption of its new standard (see Chap. 2).

The FCC has provided additional assistance as part of its new FCC Form 303-S, License Renewal. Appendix C includes three worksheets that can be used in simple situations; in essence, the FCC is just making it even easier to use OST-65. Like the earlier document, the formulas are based on ANSI-82 exposure criteria, so they would need to be revised following the adoption of the FCC's new standard. Also like OST-65, the worksheets can only be used in certain, simple circumstances: (1) there are no other towers (except for other towers within an AM array) within 315 m (about 1,000 ft), (2) there is no land any higher than the tower base within 315 m, and (3) there are not both AM and FM/TV transmitters at the same site.

All the formulas for FM and TV are based just on antenna height (above the transmitter building) and licensed power. For FM (worksheet 1), the formula assumes 100 percent downward radiation, while for TV (worksheet 3), it assumes 22 percent aural power, 20 percent downward radiation for VHF, 10 percent downward radiation for UHF, and single polarization only. Thus, if the FM station has an antenna with reduced downward radiation, the worksheet will be grossly conservative, while it would grossly *under*estimate the field for a TV station that is circularly polarized. For AM (worksheet 2), the only variable is licensed power and the only question is whether the fence is far enough out; the distance criteria in the worksheet are identical to the worst-case distances in OST-65.

Formulas for antenna arrays

The calculation of radio-frequency fields is a straightforward exercise in computation. The formulas are well known from basic physics and take the general form

$$\text{Power flux density} = \frac{\text{radiated power of RF source}}{\text{area of sphere impacted by RF source}}$$

Not surprisingly, the resulting units would be

$$\frac{\text{Power}}{\text{Area}} = \text{watts per square meter}$$

or some variant, such as the more common terms, at RF, of milliwatts per square centimeter (mW/cm^2) or microwatts per square centimeter ($\mu W/cm^2$).

The formula below comes from OST Bulletin No. 65 and is particularly relevant to the calculation of power density from intentional RF emitters:

$$\text{Power density } S = \frac{2.56 \times 1.64 \times 100 \times \text{RFF}^2 \times (0.4 \times \text{VERP} + \text{AERP})}{4\pi D^2} \quad \text{in mW/cm}^2$$

This formula can be used both for traditional TV stations, where VERP is the peak effective radiated power of the visual carrier and AERP is the effective radiated power of the aural carrier, and for analog radio and digitally encoded stations, where VERP will be zero and AERP is the total effective radiated power. The factor of 2.56 accounts for the increase in power density due to ground reflection, assuming a reflection coefficient of 1.6 (1.6 × 1.6 = 2.56). The factor of 1.64 is the gain of a half-wave dipole relative to an isotropic radiator. The factor of 100 in the numerator converts the result to the desired units of power density, when VERP and AERP are in kilowatts and D, the distance of the calculation location from the RF source, is in meters. The factor RFF is the relative field factor of the radiating antenna in the direction toward the actual point of calculation.

The formula is generally conservative due to the 1.6 reflection factor. This is intended to represent the fact that it can be assumed that there is, at the point of calculation, both the direct signal from the RF source and a secondary signal that has been reflected off the ground (or some other surface) and has also reached the point of calculation. Where such a surface is made of, say, sheet copper, the reflection factor could be 2.0 (*i.e.,* the reflected signal is as strong as the direct signal). In most cases, though, a reflected signal, if any, is well below the 60 percent assumed, which means that the calculation is assuming greater power than is really there and so is conservative. Otherwise, the formula follows basic physics and can be expected to yield accurate results.

With the common availability of desktop computers, the determination of RF power density levels by calculation is a viable means of assessing compliance with prevailing standards for limiting human exposure to RF energy. Measurements are the definitive method of evaluating RF exposure conditions, of course, but calculations have an important place, particularly in the specification of new RF facilities. By incorporating consideration of RF exposure into the design of an RF transmitting facility, antennas and antenna mounting locations can be selected to minimize RF exposure conditions.

Nevertheless, it should be noted that, while the actual calculations are quite simple, the validity of the results depends on good knowledge of operating parameters, especially 1) the radiation pattern of the RF source's transmitting antenna and 2) the elevation of accessible locations near the site. The former is obviously important since many antennas are directional in the horizontal (azimuth) plane, and virtually all are directional in the vertical (elevation) plane.

Computational programs

Most engineers working in this field have developed programs for making these calculations. The output from one such program is shown in Fig. 3.10. The first step is to develop a database of RF sources at the site and nearby. For each, the following data must be gathered: antenna location (including height above ground), type of signal (*e.g.,* TV, radio, radar), frequency of operation, radiated power, and antenna patterns. Several cautions regarding antenna patterns are in order:

1. Some manufacturers provide antenna patterns whose nulls in the elevation plane go to zero. However, unless it is a dish antenna,

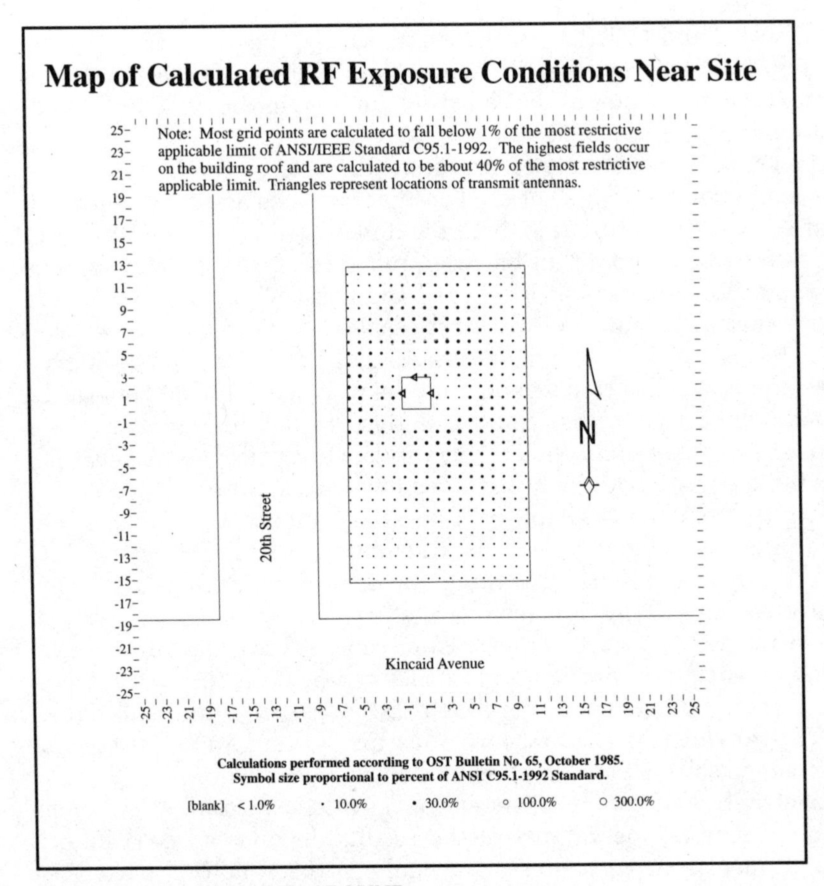

Figure 3.10 Example of RFR.GROUND output.

this is not achieved in practice, since the mere mounting of the antenna on a structure will cause some scattering, probably enough to bring the relative field pattern up to 3 to 4 percent. The program should have the ability to "fill" any pattern nulls to a specified minimum; 15 percent is a recommended level for studies addressing compliance, and some lesser value may be adequate for meeting the FCC's 1 percent exclusion criteria (see Chap. 2).

2. Manufacturers provide elevation patterns that are effective in the main beam of the azimuth pattern. That does not necessarily mean, however, that the pattern is valid at off-axis azimuths. The program should incorporate a smoothing algorithm to account for this. It is a necessary step, too, with dish antennas, where the same pattern is used in both azimuth and elevation planes. At an-

gles off either axis, dual application of the one pattern will understate the effective radiated power in that direction.

3. The FCC has expressed concern about antenna patterns provided by certain FM antenna manufacturers that show unrealistic suppression of downward radiation. Therefore, the FCC has requested that only "measured" patterns be used for RFR calculations. Clearly, this creates an inconvenience for stations applying for construction permits, as most do, before having a new antenna manufactured and measured. The FCC has indicated that it will accept analyses based upon calculated patterns derived from measured single-bay patterns, especially those characterized in a 1985 study commissioned by the EPA. In that study, single bays of five different antenna types were mounted on a tower section in a variety of ways and their resulting elevation plane patterns measured. A composite pattern for each bay type was then developed from the maximum radiation at any angle. Figure 3.11 shows a listing of these bay types, including a sixth type added in 1993, with common antenna models identified with each.

Once this database is complete, the area of interest needs to be defined. For the site terrain, a number of ground elevations can be input to the computer, which will create a smooth surface passing through them. Unless parts of the site are terraced or are otherwise very sharply contoured, this procedure gives an adequate approximation of the site; in those exceptional cases, input of additional data will cause the program to refine its calculation surface to whatever accuracy is deemed necessary. Buildings can also be defined, and the program will treat them as projections above the ground surface at whatever height is specified. Thus, nearby homes or water tanks can be modeled accurately, as well as roofs of varying heights, such as might be the case in urban areas with penthouses and elevator shafts. Without this feature, the program would have to make the assumption that the area of interest is flat, which may cause it to miss areas of increased power density levels.

Calculation methods for dish (aperture) antennas

Aperture antennas are generally round dishes that are shaped like a parabola when viewed from the side; sometimes a cylindrical extension is added for improved performance. They are used by point-to-point microwave stations, satellite transmitting and receiving antennas, radio astronomy, and radar. Aperture antennas are most often used for frequencies in the 1 to 40 GHz range and typically have gains (see Chap. 1) of 30 dB (a thousandfold power increase) or more. However, because

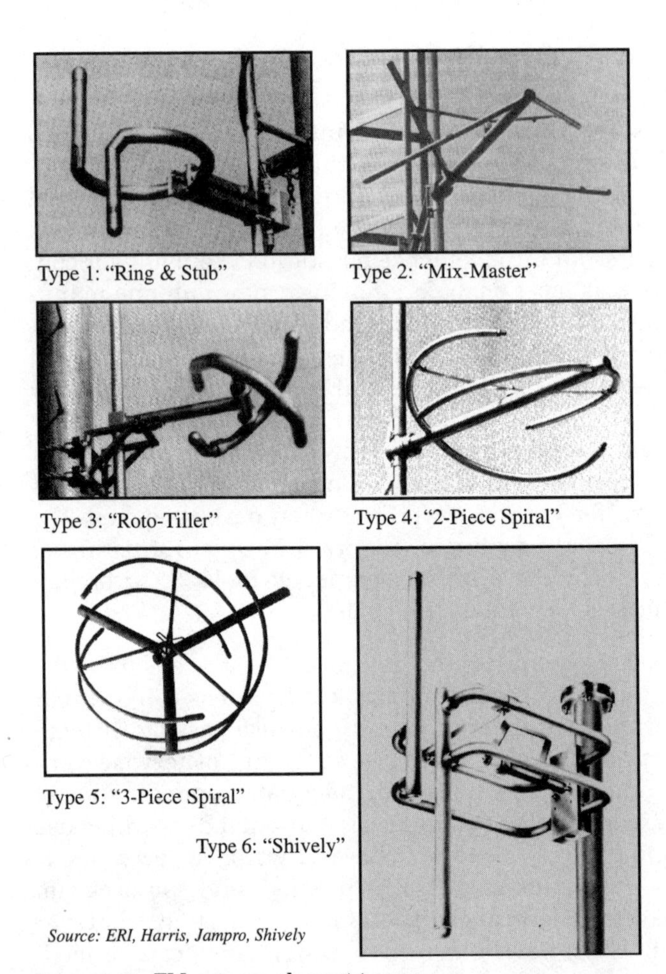

Figure 3.11 FM antenna element types.

of the highly directional nature of aperture antennas and the fact that most point-to-point microwave transmitters have low transmitter powers of 2 W or less, it is unlikely that the power density even in the main beam of the dish would exceed the PS. Further, because point-to-point microwave relay antennas are typically mounted on towers or building roofs, where human access to the main beam is improbable, the exposure levels caused by such antennas to areas of regular human access are typically far less than even 1 percent of the PS.

The same cannot be said for such uses as satellite uplink stations, where much higher transmitter powers may be involved. Although these antennas are also mounted where human access to their main

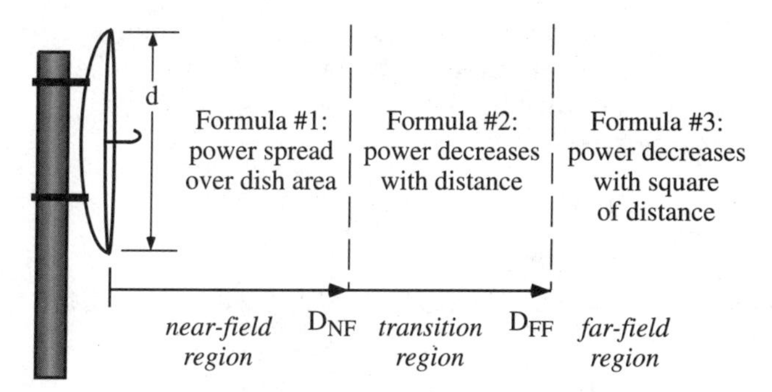

Figure 3.12 Aperture antenna calculation regions.

beams is unlikely, the power levels of the side lobes may be large enough to require calculations and, in rare instances, measurements, in order to ensure compliance with the PS.

For aperture antennas with circular cross sections, the following basic formulas can be used for first-order evaluations of their resulting power density. In virtually all cases,[15] these simple calculations will be sufficient to document compliance. There are three formulas that might apply, depending on the distance from the antenna to the point of interest at which the calculations are being done. Figure 3.12 shows the three applicable regions for an aperture antenna.

Formula 1 applies in the near field, or Fresnel[16] region, of the antenna (see Chap. 1); the power density S is given by the formula

$$S_{NF} = \frac{1.6\,\eta\,P}{\pi d^2} \qquad \text{in mW/cm}^2$$

where η is the aperture efficiency (typically 0.5 to 0.75), P is the power fed to the antenna, in watts, and d is the antenna diameter, in meters. Note that the gain of the antenna is not in the formula since the beam is not yet formed in the near-field region. This formula is valid out to a distance given by

$$D_{NF} = \frac{0.25\,d^2}{\lambda} \qquad \text{in meters}$$

where λ is the wavelength of the radiated signal, in meters. For an 8-foot dish operating at 7 GHz, for instance, D_{NF} equals 34.7 m (about 115 ft), so this formula will apply in most actual cases.

---

[15]Except for radar installations, calculations for which are a specialized case beyond the scope of this text.

[16]Named after A. J. Fresnel (1788–1827), a French physicist.

At distances beyond D_{NF}, the beam starts to form and the impressive gain figures developed by aperture antennas become a factor. The beam is fully formed in the region known as the far field, or Fraunhofer[17] region, defined by distances exceeding

$$D_{FF} = \frac{0.6\, d^2}{\lambda} \qquad \text{in meters}$$

For the example above, D_{FF} equals 83.2 m (about 275 ft), and formula 3 applies:

$$S_{FF} = \frac{P\,G}{40\,\pi\,D^2} \qquad \text{in mW/cm}^2$$

where G is the antenna power gain, relative to an isotropic radiator, and D is the distance to the calculation point, in meters. This formula reflects the inverse square law; thus, while the full value of the gain is now applied, the distance is so large that the power density can be quite low.

At distances between D_{NF} and D_{FF}, known as the transition or crossover region, formula 2 gives the power density:

$$S_{TR} = S_{NF}\,\frac{D_{NF}}{D} \qquad \text{in mW/cm}^2$$

Note that D will always be greater than D_{NF}, since this is the transition or crossover region beyond the near field, so S_{TR} will always be less than S_{NF}, as would be expected.

On-tower exposure calculations

Compliance with the FCC limits on human exposure to RFR requires consideration of the exposure to persons working on the towers that support radio communications antennas. Since close approaches to or direct contact with broadcast antennas may be involved in such situations, it is necessary to reduce the power being radiated from those and other nearby antennas, sometimes even to cease operations, while workers are on the tower. Obviously, the stations involved will wish to operate in this condition for as short a time as possible, and it is therefore necessary to determine, for a specific height on each tower at a site, what reduced-power conditions are required.

An accurate calculation model is the preferred method for developing the appropriate power reduction schedules to meet this need.

---

[17]Named after Joseph von Fraunhofer (1787–1826), a German optician and physicist.

Certainly, it is possible to develop such schedules by actual measurements, and such findings would be more definitive than calculations. There are several logistical problems, however, with the measurement approach, as noted below:

1. Each tower at the site must be climbed, and each station at the site needs to be ready, upon request, to reduce its power for extended periods, including the possibility that it might have to shut down (presumably during daylight hours). Positive communications with each station is vitally important for the safety of the rigger, so additional staff may be required for monitoring and recording, and "lock-out" protocols should be followed. Problems can develop with two-way radios or cellular phones, which may be overloaded by the high fields and become unreliable.
2. A tower rigger needs to be educated in the measurement techniques for the meter(s) being used, and care needs to be taken that the rigger is not exposed to fields in excess of ANSI.
3. The process will be excruciatingly iterative: climb until the PS limit is reached, reduce power on those station(s) suspected of being the major contributors, climb until the limit is reached again, reduce power some more, climb, reduce, and so on.
4. It may not be clear just when stations can turn their power back up (*i.e.,* when the rigger has climbed *past* the area of high fields). If this is tested, the risk of exposing the rigger is increased, especially when he or she needs to climb back down.
5. Days or even weeks could be required to complete the process at even a small antenna farm.
6. The power reduction schedules developed may be of no value after any change, such as the addition of a new RF source or the replacement of an antenna, is made at the site.

For consideration of ground-level exposures, one can legitimately apply far-field assumptions and treat the antennas as point sources (*i.e.,* all energy radiating from the center of the antenna). However, for tower work, that assumption cannot be made, and the rigger is often in the near-field of an antenna on the tower being climbed or on a nearby tower. To account for this, the aperture of each antenna should be included in the database, in addition to the data required for the ground-level calculations. The program can then make certain conservative assumptions:

1. When calculating power density levels on that tower, the power is spread over the aperture of the antenna. This is analogous to the

power being divided among a discrete number of elements (bays), as is the case with most FM and TV antennas.

2. When calculating power density levels on a nearby tower, the power is concentrated at the height within the antenna aperture that most closely equals the calculation height. In this way, the power is assumed to be a point source at the bottom of the antenna as the rigger climbs toward it, then "follows" the rigger up the tower, and finally becomes a point source at the top of the antenna once he or she has climbed above it.

A good program will include features to make it easy to test various conditions, and this is a major advantage over a measurement program. For instance, "what if" tests can be made to develop power reduction schedules that involve the least number of stations or are otherwise easiest to implement. The results should then be presented as a table, by tower section, of appropriate power reductions, as shown in Fig. 3.13.

The one limitation to the assessment by calculation of on-tower exposure conditions is that localized hot spots cannot be predicted. This factor should be evaluated by measurement the first time that a tower is accessed in accord with the power reduction schedules developed by calculation.

Schedule of Recommended Operating Powers to Achieve Calculated Compliance with ANSI C95.1-1982

Tower A: Height on Tower (ft)	*Ch. 22* WABC	*97.9* WDEF	*102.3* WGHI main	*102.3* WGHI *aux*
80–top	0	25	50	*100*
60–80	0	25	50	*100*
40–60	100	25	25	*100*
28–40	100	50	50	*100*
0–28	100	100	100	*100*

ANSI Height: 28 feet

Notes: Calculations based on FCC OST Bulletin No. 65, October 1985.
Entries in table represent recommended percentage of licensed power.
Entries in italics indicate alternate operation on auxiliary facilities (if available).
Power reductions apply for any access of any duration to the pertinent tower section.
Stations not appearing in table may operate at full power on either main or auxiliary facilities.

Figure 3.13 Example of RFR.TOWER output.

Mitigation in Uncontrolled Environments

Whether based on measurements of RF power density levels or based on their prediction by calculations, if fields exceeding the PS limits are indicated, mitigation is a required step.[18] Mitigation can consist *only* of the following:

1. Reducing the field below the PS
2. Restricting access to the fields exceeding the PS
3. Some combination of the two

Compliance can be achieved via different types of mitigation measures for controlled (occupational) or public (uncontrolled) exposure situations, since it is presumed that workers can be taught to abide by access restrictions or operational procedures. The goal of all mitigation measures is to ensure that the PS is, beyond reasonable doubt, being met. Because this is an area of public safety, and because of the all-too-litigious nature of contemporary American society, that goal should be achieved with enough conservatism that one can be *certain* that the PS is being met, regardless of who is checking for that condition. The PS provides a presumptive safe harbor under the law, which an entity responsible for creating RF emissions should certainly seek. For instance, an employee who smokes two packs of cigarettes a day might argue at some time in the future that he or she got lung cancer from "all that RF" against which the employer had failed to provide adequate protection.

For uncontrolled settings, no training level should be assumed. The strong preference for mitigation is option 1 above. Options 2 and 3 are acceptable alternatives, but the barriers against access should be so complete that no special notification is required, and no literacy level should be assumed, either.

FCC requirements

The FCC, as the body licensing intentional emitters of RF, in 1986 provided guidance[19] for the broadcast industry on achieving compliance with ANSI-82, its adopted standard at that time. With prescient distinction, seven different situations were identified, (A) through (G) as

---

[18]Even if the FCC, or any other body, does not require it, or enforce it, prudent management practices do both.

[19]Federal Communications Commission, "Further Guidance for Broadcaster Regarding Radiofrequency Radiation and the Environment," Public Notice, January 28, 1986.

(A) High RF levels are produced at one or more locations above ground level on an applicant's tower
- If the tower is marked by appropriate warning signs, the applicant may assume that there is no significant effect on the human environment with regard to exposure of the general public.

(B) High RF levels are produced at ground level in a remote area not likely to be visited by the public
- If the area of concern is marked by appropriate warning signs, an applicant may assume that there is no significant effect on the human environment with regard to exposure of the general public. It is recommended that fences also be used where feasible.

(C) High RF levels are produced at ground level in an area which could reasonably be expected to be used by the public (including trespassers)
- If the area of concern is fenced and marked by appropriate warning signs, an applicant can assume that there is no significant effect on the human environment with regard to exposure of the general public.

(D) High RF levels are produced at ground level in an area which is used or is likely to be used by people and to which the applicant cannot or does not restrict access
- The applicant must submit an environmental assessment. This situation may require a modification of the facilities to reduce exposure or could lead to a denial of the application.

(E) High RF levels are produced in occupied structures, on balconies, or on rooftops used for recreational or commercial purposes
- The applicant must submit an environmental assessment. The circumstances may require a modification of the broadcasting facility to reduce exposure or could lead to a denial of the application.

(F) High RF levels are produced in offices, studios, workshops, parking lots or other areas used regularly by station employees
- The applicant must submit an environmental assessment. This situation may require a modification of the facilities to reduce exposure or the application may be denied. This situation is essentially the same as (E). We have included it to emphasize the point that stain employees as well as the general public must be protected from high RF levels. Legal releases signed by employees willing to accept high exposure levels are not acceptable and may not be used in lieu of corrective measures.

(G) High RF levels are produced in areas where intermittent maintenance and repair work must be performed by station employees or others
- ANSI guidelines also apply to workers engaged in maintenance and repair. As long as these workers will be protected from exposure to levels exceeding ANSI guidelines, no environmental assessment is needed. Unless required by the Commission, information about the manner in which such activities are protected need not be filed. If protection is not to be provided, the applicant must submit an environmental assessment. The circumstances may require corrective action to reduce exposure or the application may be denied. Legal releases signed by employees willing to accept high exposure levels are not acceptable and may not be use in lieu of corrective measures.

Figure 3.14 FCC exposure situation guidelines.

listed in Fig. 3.14, each with a suggested mitigation protocol, and those basic situations are relevant even when the evaluation is made against the more current ANSI-92 or some other standard. Situations (A) through (F) represent uncontrolled exposure situations [even situation (F) where the parking lots and offices of a broadcast station are probably visited frequently by salespeople, office equipment maintenance staff, and janitorial services, none of whom can be expected to have been educated about the particular environmental factors present].

In practice, of course, the suggested environmental assessments are not made, since the FCC (or any other review body) cannot be expected to condone an operation *not* meeting the PS guidelines. Thus, the high fields in those areas of public access should be removed by raising the offending antenna(s), changing to antenna(s) that have less downward radiation, or relocating the station(s). Failing that, the area of high fields should be *secured* against public access; signs, for instance, would not be adequate to provide assurance against exposures in excess of the PS. Instead, the areas of concern should be fenced or otherwise enclosed and points of access should be locked. It is interesting to note the FCC's suggestion that even trespassers need to be protected. This means that gates in fences must be securely locked and that fences, when relied upon to achieve access restriction, should be topped with barbed or razor wire. The latter material, being stamped from a single strip of metal, is preferred for avoiding the generation of intermodulation products at sites with numerous RF sources.

Warning signs

ANSI Standard C95.2-1981 (reaffirmed in 1989) specifies the composition and coloring of RF warning signs, as shown in Fig. 3.15. Such

Figure 3.15 Standard RF warning sign.

signs are available from a variety of sources.[20] Warning signs incorporating this information are recommended, *as an extra precaution,* on secure fences and locked doors. It is important to recognize that signs are not sufficient in and of themselves in areas of public access, since it cannot be assumed that the persons being protected can read [*e.g.,* they may be too young to read or they may not read the language(s) used on the sign]. Especially with the FCC's comments about trespassers, there are few if any situations in which signs alone are adequate to mitigate possible high exposures in an uncontrolled environment.

Time averaging

Time averaging should not be relied upon for ensuring compliance with the PS for uncontrolled areas, subject to public access. The applicable time-averaging provision on the current standards is 30 min. However, the time limit is useful for mitigating exposure to high fields only if exposure times will be considerably *less* than 30 min. For example, fields twice the limit would require that any exposure duration be limited to 15 min, *and* that the 15 min both before and after exposure have zero exposure. In certain controlled environments, discussed in the next section of this chapter, time averaging can be an effective mitigation measure. However, since access to the area of concern here is, by definition, *un*controlled, one cannot determine with any certainty how long a member of the public would remain in that area.

Perhaps for radios on a ferry boat whose route is a short one, the time-averaging provision could be justified, but even then what is to prevent someone from riding over and back? As a practical matter, time averaging provides no benefit in developing a mitigation program for high fields in an uncontrolled environment.

Transient exemption

Both ANSI-92 and NCRP-86, the two principal U.S. two-tier exposure standards, contain exemptions from the tighter public guidelines under the limited circumstance of short-term exposure. These are *not* exemptions from the less restrictive occupational guidelines, which cannot be exceeded. Since both ANSI-92 and NCRP-86 already allow for 30-min time averaging, the exemptions must be intended to apply for exposures of a longer duration. ANSI-92 describes the circumstance simply, as follows: "transient passage through areas where analysis shows the exposure levels may be above [the uncontrolled

---

[20]Sources include the National Association of Broadcasters, Washington, D.C. 202/429-5300.

limits] but do not exceed [the controlled limits]." This phrase is footnoted at the two locations it appears in the Standard, as follows: "The means for the identification of these areas is at the discretion of the operator of a source."

Does this mean that a radio station licensee can declare, for instance, that the parking lot near the radio tower is such an area of "transient passage" and therefore avoid the need to mitigate fields not exceeding the controlled limits? There is nothing in ANSI-92 to suggest that this cannot be done, although that would appear to contradict the FCC's expectations, such as those expressed in situation (F) in Fig. 3.14, as well as the whole concept justifying a second tier for uncontrolled environments. ANSI-92 defines uncontrolled environments as those "locations where there is the exposure of individuals who have no knowledge *or* control of their exposure [emphasis added]." It would appear that the parking lot would be an uncontrolled environment, in which office employees might linger, say, at picnic tables, longer than 30 min. While they may have knowledge of the exposure conditions, they have no control over them; hence the uncontrolled designation.

If it is argued that the office employees do have control, since their presence there was voluntary, the same logic could be applied to a residential setting, "where control of their exposure" would include the ability to move away. Or, if it is argued that both knowledge *and* control must be absent to constitute an uncontrolled environment, notification of employees or nearby residents would be sufficient to turn an uncontrolled environment into a controlled one. As a conservative approach, either lack of knowledge or lack of control *of the RF levels* (rather than literal control of one's presence) should be sufficient to trigger the uncontrolled definition, and use of the "transient passage" exemption should be made only in rare circumstances.

NCRP has delineated a clearer situation, specifically suggesting circumstances that can not be anticipated and should not recur:

> It is recognized that there are special circumstances in which the exposure limits for the general population may unnecessarily inhibit activities that are brief and non-repetitive. For example, the presence nearby of a number of emergency vehicles engaged in telecommunications might cause a brief exposure to fields at strengths above the general-population limits. Because only small groups of the population would be exposed under these conditions, and almost certainly not on a repeated basis, the occupational exposure levels are permitted for such cases.

If ANSI had this situation in mind for its exemption, it is unfortunate that this clear language from NCRP's work was not appropriated. Nevertheless, as a practical matter, neither standard provides any leniency that should be relied upon in developing a mitigation program for an ongoing situation.

Tower security

Fundamental to achieving compliance with the public exposure limits of the PS at a telecommunications site is restriction against unauthorized climbing of towers. At one site, it is an annual prank for fraternity pledges from the local college to climb the 400-ft TV tower and steal the aeronautical beacons, despite heroic efforts by the TV station to secure the base of the tower. While most trespassers are not so dedicated (or foolish), a tower could be considered an attractive nuisance and, as such, must be maintained so as to prevent at least casual climbing by persons at the site. Removal of lower sections of ladders should be done, at a minimum; preferably, the base of each tower or each cluster of towers would be fenced with 6-ft cyclone fencing, topped with razor wire and locked at all gates. Finally, of course, the potential for exceeding the PS exists, as well, at height on the tower, and standard RF warning signs should also be mounted where they would be visible to unauthorized persons attempting to climb the tower.

Mitigation in Controlled Environments

The mitigation of high fields in occupational settings is different from that reviewed in the previous section. A controlled environment is defined by ANSI-92 as those "locations where there is exposure that may be incurred by persons who are aware of the potential for exposure as a concomitant of employment [or] by other cognizant persons."[21,22]

Controlled environments can also be defined by what they are not (*i.e.,* uncontrolled: "persons who have no knowledge or control of their exposure"). Therefore, in a controlled setting, it can be assumed when developing a mitigation program that the people involved can be instructed in certain procedures to be followed in certain situations. This allows considerably more latitude in what would be considered acceptable mitigation measures.

Nevertheless, it is important that the entity causing the high fields take active responsibility for that mitigation. The FCC has indicated that this responsibility cannot be signed away, meaning that people who might be exposed to high fields cannot waive the provisions of the

---

[21]*Concomitant* is defined as "existing or occurring with something else, often in a lesser way."

[22]The troublesome "transient passage" part of the definition, including areas that would otherwise be uncontrolled, has been reviewed in this chapter.

PS; the courts would probably agree. A review of present-day practices at transmitting sites around the country would reveal that many broadcast stations rely on the tower riggers themselves to meet the PS working conditions in high RF fields and that, unfortunately, many riggers exhibit a cavalier attitude toward compliance with the PS.

When calculations are used to define when mitigation might be required, a 50 percent threshold is recommended, by which all fields calculated to be 50 percent or more of the PS would be mitigated. In most cases, if the calculations are this high, measurements will be made to confirm the actual 100 percent contour location, as well as to check for hot spots. This rule of thumb applies for ground-level exposure conditions, where the 1.6 ground reflection factor suggested by OST-65 is appropriate (which increases the calculated power density by a factor of $1.6 \times 1.6 = 2.56$). For on-tower exposure conditions, the power reflected by the ground is less significant; when the same equation is used, the mitigation threshold can be raised to 100 percent.

OEG

The occupational mitigation practices developed for a site are often summarized in an Occupational Exposure Guide, or OEG.[23] This document is generally produced as a two-sided laminate approximately 8½ by 11 in in size. The front side would have (1) a short description of the PS, (2) an explanation of the significance of the various mitigation markings at the site (*e.g.,* orange stripes and signs), and (3) a telephone or pager number to contact for further information. The back side would have the power reduction tables for on-tower access. These laminates would be produced in sufficient quantity that one could be hung on the inside of every outside door at the site and could be kept at each station's office. In this way, everyone permitted to be at the site should have adequate knowledge of the RF environment as determined at the time the OEG was developed. Before non-RF workers (such as roofers, HVAC repair persons, and utility meter readers) are authorized access to the site, each should be provided with a copy of the OEG, offered the chance to ask questions about it, and required to sign an acknowledgment, including the commitment to abide by it.

Work that involves tall, temporary structures being brought to the site should be reviewed beforehand for the possibility of creating potential hazards. For instance, if a small truck-mounted crane were used to hoist a replacement air conditioning unit onto the roof of a transmitter building, the potential for RF shock could be created at ground level.

---

[23]A term coined by the author in 1986 and now in use at a number of antenna farms.

Similarly, if a bucket truck were used by the utility company to bring in new electrical service on poles, workers might be put into high fields that, being neither at ground level nor on a particular tower, had not been assessed. The measurement and/or calculation techniques already described can be used to evaluate such situations; the important issue is that they do need to be studied beforehand so that appropriate mitigation measures can be developed and implemented.

Field markings

Clear identification of areas subject to high RF power density levels is essential to the effectiveness of the mitigation program. (The assumption is made here that public access has already been precluded at a greater distance and that the concern now is adequate communication with persons authorized to be within the areas exceeding the public exposure limits.) There are several guidelines that good engineering practice suggests be followed:

1. Identification should be done in a manner that will last for many years without high levels of maintenance.
2. Distinctive coloration should be used. The author is partial to International Orange; although the yellow in ANSI C95.2 can be effective, particularly in combination with black; and bright red markings have been highly satisfactory, especially in light of the intuitive American "halt" response to that color. Whatever color is selected, it should be used consistently; for instance, orange in some areas and red in others would be confusing. If a second color is needed, to mark different contour levels, for instance, it should be contrasting; both green and blue should work, and each one can be obtained in the bright hues needed for high visibility.
3. Markings should be continuous or at least easily followed by eye around the area of concern. A painted post at each end of a 30 ft area, for instance, would not normally be adequate. As shown in Fig. 3.16, which illustrates many of these suggestions, a band of color can be used effectively. This band, at a constant height (or width) of perhaps 6 in, is painted on the sidewalk, painted up the side of the building to perhaps 4 ft above ground, and follows the face of the building until the point where the contour resumes its march across the grounds. Posts[24] are put in the ground about every 5 or 6 ft around the contour; they are as tall as the stripe on

[24]Ideally, the posts would be made of dielectric materials, such as concrete-filled PVC, so as not to create new localized hot spots. The necessity for this largely depends on the measurement distance set forth in the PS.

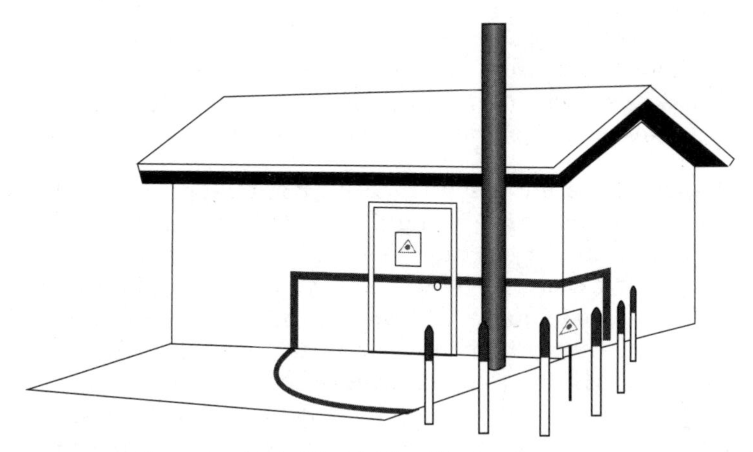

Figure 3.16 Example of high RF field identification.

the building and have the same band of color on top. (Shorter posts have been used at certain sites; they should be placed much more closely together than the taller posts in order to maintain the line of visual continuity.) This continuous marking method has the advantage of being visible and easily interpreted from any angle of approach. (Note that the pole shown *inside* the contour is not striped; the concept of a boundary is easily recognized and a similar marking in the middle would be confusing.)

4. Signs must be large enough that they can be read easily from *outside* the areas of concern. The meaning of the message should be apparent upon initial sighting; long, involved messages or instructions should be addenda to the OEG and placed elsewhere. Use of the standard C95.2 yellow triangle and RF radiation symbol is not essential on signs used to mitigate occupational exposures, although it may be desirable. The purpose of the signs in this setting is less to alert authorized persons to the presence of high fields than it is to provide guidance about the actions to take as a result. For instance, the sign in Fig. 3.16 might read

WARNING

RF levels exceed ANSI.
Do not remain
within marked area
more than 2 MINUTES.
See OEG for more information.

5. Conditional requirements for access should have readily available indications of status. For instance, if an area has high fields only

when certain antennas are operating,[25] an indicator light could be installed to alert personnel that those antennas are in use. (A means to check operation of the light might be a good addition, too, with contact information provided in case the light is not working.)

Localized hot spots present a special case for mitigation since individual paint stripes would be confusing. The best approach is actually to fill much of the volume of the high field and so preclude access, in which case no marking would be needed. Many hot spots are associated with guy wires; the foam material made for insulating water pipes, available at any hardware or home improvement store, is particularly useful for this purpose. Larger spacers can be made of wood or fiberglass. The alternative is a sign at every hot spot or group of hot spots not filled, alerting people that the localized high field exists and that they should stay away.

Personal monitors

Personal radiation monitors, described in more detail earlier in this chapter, can be an effective part of a mitigation program for occupational exposure conditions. Localized fields are too difficult to predict by calculation, and they may change size, location, and magnitude as a result of work that might be taking place at a site. For instance, a technician installing a new microwave link would be handling dishes and feedlines that could create localized fields as they are moved up to and into position. Presumably, this work is being done in compliance with the power cutback schedule developed for the height on the tower at which work is being done, but fields in excess of the guidelines could be created even in those circumstances.

Therefore, use of personal monitors for such workers is recommended, as a matter of routine. They would be alerted whenever the exposure conditions reached 50 percent of the PS, at which time they should secure their work, remove themselves from the field, and arrange for the fields to be reduced. The personal monitors should be worn in accordance with the manufacturer's recommendation (*i.e.*, on the chest outside of any clothing). Daily protocol should include activation of the self-test feature, if available, or test activation using a hand-held communications device or other active RF source.

---

[25]This is a frequent situation at antenna farms, where standby, backup antennas will be mounted lower on the towers and can have a much larger contribution to the RF levels, when in use.

Spatial averaging

Spatial averaging at the limit of ANSI exposure guidelines is a poor engineering practice, especially when metering uncertainties (discussed earlier) have not been taken into account. Averaging will give indications that are not conservative and may well conceal high fields for which mitigation is required.

Spatial averaging is sometimes claimed as justification for not mitigating known localized fields whose power density exceeds the occupational limits. However, that allowance is a feature only in ANSI-92, not in NCRP-86, and ANSI-92 effectively takes that option away, itself, with its exclusion against high localized fields that might affect one's eyes or testes. These organs were identified because of their relative lack of ability to dissipate heat. That exclusion cannot be ignored; the situation is rare in which a localized field is situated in just the right way so that no one could place either eyes or testes in it. Therefore, a conservative mitigation program will not rely upon spatial averaging of localized hot spots.

Power cutback tables

At a site with more than one RF source, consideration has presumably been given to the on-tower exposure conditions, with one RF source possibly illuminating the area around another. The power cutback tables developed as a result should be an integral part of the mitigation program developed for the site. It is too difficult to predict by calculation the presence of hot spots, so this issue needs to be addressed via measurement, should repeated or prolonged access to a particular tower height be required. Accordingly, one of the first steps in a tower rigging project is to measure the local RF conditions at the height at which work is to be done.

It is essential that positive control be asserted over RF sources whose power reductions are necessary for tower access. This should be accomplished through the OSHA "lockout/tagout" protocol, whereby the power control is tagged with a sign while the reduction is in force so that someone does not inadvertently turn the power back on or up while the worker is still on the tower in the area of concern. Preferably, the switch would be locked with a padlock in its OFF position, although the power cutback tables have probably been developed with the intent of keeping as many stations on the air as possible, so a physical lock on the power to the transmitter is not always an option. Of particular importance is disabling the remote control capability while on site so that station personnel at a distant location, who might not be aware of the tower activity, do not have the ability, having seen the low-power condition, to raise it back up.

Time averaging

Both ANSI-92 and NCRP-86, the two most recent American standards, allow averaging of exposures to high fields over some unit of time. For occupational exposure situations, and as part of a comprehensive mitigation plan, this capability can provide workable solutions for accessing or traversing areas where fields exceed the standards. These standards allow averaging over a period of 0.1 h (6 min),[26] with the average not to exceed the limit for continuous exposure.

Correct time averaging of exposure conditions considers the time both *before* and *after* the period of high exposure. Thus, for a 2-min exposure to high fields, a 10-min period must be evaluated so that the first 6 min (4 out and 2 in) and the last 6 min (2 in and 4 out) both pass the test. Also, the level of exposure when "out" of the high field is equally significant; an exposure to a field 3 times the limit could be accommodated for 2 min *only* if the 4 min before and the 4 min after have *zero* exposure. In practice, of course, that is not always the case, so access times for an area of high fields are generally based, conservatively, upon the maximum accessible field within the area, and the area is defined to be that within which the fields exceed some *fraction* of the PS. In this way, it can be assumed that the time "out" will not exceed that fraction of the PS, and the allowed time "in" can be determined from the following formula:

$$\frac{\max_{in} \times \text{time}_{in} + \max_{out} \times (6 - \text{time}_{in})}{6} = 1$$

which simplifies to:

$$\text{time}_{in} = 6 \times \frac{1 - \max_{out}}{\max_{in} - \max_{out}} \qquad \text{in min}$$

If the area of concern is defined by fields exceeding 50 percent of the PS, $\max_{out}$ equals 50 percent and the formula becomes

$$\text{time}_{in} = \frac{3}{\max_{in} - 0.5} \qquad \text{in min}$$

---

[26]ANSI-92's reduction in this time period begins at 15 GHz and so has limited practical impact.

Thus, for a maximum field of, say, 3 times the PS, an access time "in" of 3 ÷ 2.5 = 1.2 minutes would be allowed.

Clearly, the lower the PS fraction used for the area definition, the larger that area will be, but the longer the allowed time "in" will be, too. Trade-offs, of course, will need to be made in each particular case. This does mean that a worker must actually move away from the field during the periods "out." Therefore, this mitigation measure will not work for all situations, and power reductions may be required for extended work periods. However, for such activities as reading meters or crossing to another area of low fields, time averaging may be used effectively as a mitigation for high fields in controlled areas.

It should be noted that the time-averaging definition given by ANSI-92 itself is so simplistic as to be strictly incorrect. Within that definition is the formula for determining relaxed exposure levels during some known time of exposure T_{exp} less than the time period T_{avg} specified in the guidelines:

$$\text{MPE}' = \text{MPE}\,\frac{T_{avg}}{T_{exp}}$$

This formula does not account for the remainder of time ($T_{avg} - T_{exp}$), either before or after, and does not account for exposure levels during those bracketing periods. In the example above, an allowed T_{exp} equals 2 min, which exceeds the correct answer of 1.2 min by a considerable margin (67 percent) and would yield a time-averaged exposure exceeding the guidelines. The ANSI formula is correct only for those limited cases in which T_{exp} recurs no more often than at intervals of T_{avg} and in which all exposure levels are zero at times other than T_{exp}. Because these limiting conditions are not normally met, the ANSI formula should not be used in practical applications.

RF radiation suits

Increasingly popular among tower riggers are coveralls made of a material designed to attenuate RF energy. However, as discussed below, such suits generally should *not* be relied upon for mitigating conditions of high RF power density.

Although several manufacturers have produced "radiation suits" over the years, there is only one company presently making the material and the suits: NSP Safety Products.[27] The material, called Naptex,

[27]NSP Sicherheits Produkte GmbH, Haupstraße 17 D-86695, Nordendorf, Germany, telephone 08273/1031.

is cotton cloth interwoven with polyester/stainless steel threads. Two grades were manufactured recently, PM20, with alternating Naptex and cotton threads, and PM30, with only Naptex threads. The PM20 material was light blue and the PM30 olive green, although PM30 is now manufactured in "safety orange," and this color is presently used for the construction of suits marketed in North America.

The material has been extensively tested, and its effectiveness at attenuating RF energy is well characterized. Literature from one of the U.S. distributors indicates that PM30 provides attenuation of 30 to 40 dB up to about 15 GHz, falling to about zero at 100 GHz. Fewer studies have been performed on the material when it has been configured as a suit. Available data shows that the suit and hood, when worn without the optional "overshoes," can actually *exacerbate* body current. This effect is shown in Fig. 3.17, which summarizes data from testing funded by the manufacturer. The presumption is that the suit is a more effective receiving antenna at some frequencies than is the body, and that the energy thus received is then coupled to the body, which must conduct it to ground. With the overshoes in place, the energy is conducted to ground directly, without passing through the body, which is why the current to ground (called "body current" by the researchers) remains high while the SAR at the ankle drops markedly.

As part of its research under the Docket 93-62 proceedings, in 1995 the FCC commissioned a study (RFP #94-92) of several aspects of the technical issues under consideration. With regard to the suits, the study reached much the same conclusion as that above: "The available data seem to support the concept that because of the internal conduc-

frequency	*test condition*	*body current*	*ankle SAR*
2.03 MHz	no suit	213 mA	2.4 W/kg
	suit & hood	256	2.7
	plus overshoes	115	0.0
29.9 MHz	no suit	598 mA	15.5 W/kg
	suit & hood	840	15.7
	plus overshoes	674	1.1
80.0 MHz	no suit	640 mA	19.1 W/kg
	suit & hood	700	23.7
	plus overshoes	740	0.7
400 MHz	no suit	122 mA	2.2 W/kg
	suit & hood	(no data reported)	
	plus overshoes	110	0.0

Source: Olsen & Van Matre, 1993

Figure 3.17 Ankle SAR reduction.

tive quality of the Naptex suit, the suit itself, without an appropriate pathway to ground, will lead to...slightly enhanced body currents as measured via foot current meters or ankle SAR measurements."

The problem with these findings is that the overshoes, which are cloth booties made of the Naptex material, are not practical. Tower work is hard on shoes, since the climbing surfaces are normally steel, which may even have been roughened for traction. A thin layer of fabric on the bottom of the shoes, even if not worn while on the ground, would still be destroyed in short order. Most riggers would not even put them on, though, due to the potential safety issue raised by the poor traction of Naptex cloth relative to rubber- or leather-soled shoes.

The FCC study speculated about the possible benefit of Naptex socks but acknowledged that no testing of that option has been done. In concept, the sock, if bonded to the suit itself, would provide a complete shield to the lower extremities, so that energy absorbed by the suit need not be coupled to the body for conduction to ground. It is important that the socks be tall enough and loose enough so that they fit *over* the worker's pant leg and make several inches of contact with the legs of the RF suit. Some way of securing the pant leg at that location, such as a Velcro strap, would be an effective means. Tests have shown considerable isolation of the body from ground by normal shoe and boot soles; this data suggests that Naptex socks may not be as effective as, say, a conductive work boot, although considerable benefit would be expected. The author has been encouraging NSP since 1993 to develop a conductive boot, and the company now has a prototype, which was shown at the 1996 National Association of Broadcasters Engineering Conference. With this type of boot bonded properly to the body of the suit, there is no reason any longer for the energy to couple to the body.

The hood has also caused usage problems. Because it is not tight fitting, which would be uncomfortable, the hood tends to stay in place when the rigger's head is turned, and the rigger ends up looking at cloth in that situation. Having a field of vision thus limited to, perhaps, $\pm 45°$ presents a safety concern for riggers working on a tower and handling materiel being hauled up on cables; the hood also makes photographic work more difficult. Without the hood, of course, an important part of the body is unprotected, which is why the present version of the suit has the hood sewn on, so at least the rigger cannot leave it at the tower base. Still, it is questionable how many riggers pull the hood up, and how many simply let it hang behind them. Newer versions of the suit have Velcro inside the hood, with the idea that the wearers of the suit would attach mating Velcro pieces to their hardhats, thus moving the hood, too, when they turn their heads.

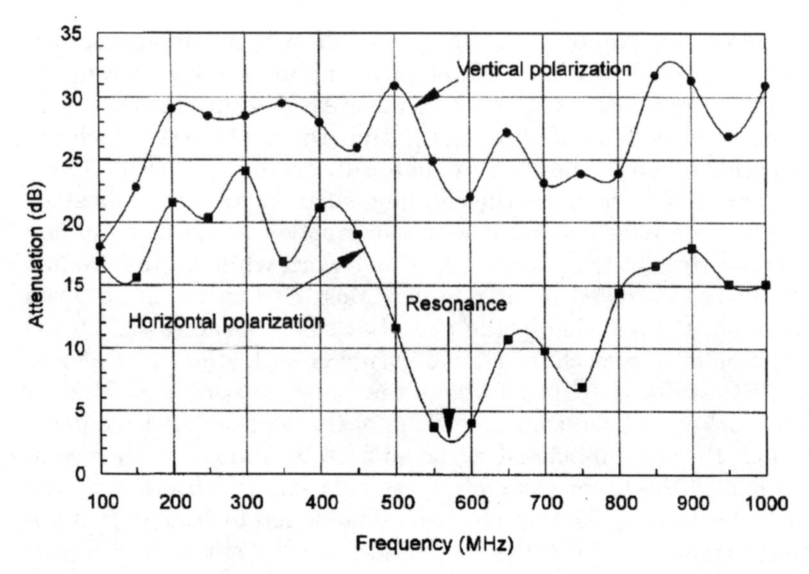

Source: 1995 FCC Study (RFP #94-92)

Figure 3.18 German Telecom Naptex suit measurements.

Gloves made of Naptex are also available and are listed in the distributor's literature as "required," with the expectation that they would be worn beneath typical leather gloves. It has been reported that care must be taken to make and maintain a secure joint between the gloves and sleeves of the suit in order to avoid arcing across a gap at that location when the suit is worn in high fields.

The findings reported above on the value of providing a path to ground are important, as are the findings of German Telecom, which has done some of the most extensive testing of the suit as a whole. Figure 3.18 is taken from the 1995 FCC study and shows the results of the German studies. Almost all U.S. FM and TV stations broadcast a horizontally polarized component, and the tests identified a resonant condition in the middle of the UHF band for horizontally polarized radiation. Data at frequency increments less than 50 MHz may show the presence of other resonant conditions, as well. It is believed that the resonance may derive from the front-to-back or side-to-side dimension of the suit, assuming a propagation path directly toward an erect figure; if so, that means the condition of resonance would vary in frequency, depending on suit size, rigger's girth, and even body position. Note, too, that the depth of the curve is not defined by experimental data; the actual attenuation at the resonant frequency could be less than the 3 dB shown (and could even be negative).

Thus, one might concentrate on use of the suit at sites with only VHF-TV and FM frequencies, where it would appear to provide attenuation not subject to resonance peaks. Unfortunately, the data shown in Fig. 3.18, as with most of that from other sources, does not cover frequencies below 100 MHz. As further work is done, it will be of particular interest to review similar data on performance of the suit when long-axis resonance develops, which would be expected to occur below 100 MHz.

Practical and effective RF suits would certainly be a great boon to the industry, and it is hoped that research and design refinement in the years ahead will provide solutions to the concerns identified here and lead an agency such as FCC or OSHA to approve their use in practical settings.

Chapter

4

Media's Role

Prominent Media Coverage

Brodeur

By now, virtually every major metropolitan newspaper (and many smaller ones) has had articles, sometimes running on the front page, about both ELF and RF radiation, its increasing presence, and its purported health effects. The journalist pioneer, however, was Paul Brodeur,[1] writing in the 1970s in *The New Yorker* magazine. It was Brodeur who introduced into public consciousness the idea that there might be something harmful in, primarily, high-voltage power lines and, secondarily, radio communication facilities.

Brodeur, a novelist and investigative journalist, apparently believed he was breaking a scandal in public health and industrial safety. His several investigative articles in *The New Yorker*, later turned into books, have been positioned as daring exposés of severe public and occupational health risks of electromagnetic fields and, more importantly, of a pervasive conspiracy between industry and government to cover up information that could have affected public safety. He was, perhaps, overly sensitized to conspiracies by his service in the U.S. Army Counter Intelligence Corps (1953–1956). In his zealousness, however, he tends toward alarmist and inflammatory rhetoric, playing on people's innate fears of what they do not understand and relying primarily on anecdotes, rather than evidence, to support his theories.

Although well-received as a novelist, Brodeur has been faulted by literary reviewers of his nonfiction writings for his bias, his quickness

---

[1]Paul Adrian Brodeur, Jr., born 1931, earned his bachelor's degree at Harvard.

to rely on conspiracy theories, and his failure at times to explain clearly the scope, limitations, and meaning of the evidence he uses to support the claims he makes. These criticisms apply to both of Brodeur's books on ELF and RF fields. His first book, *The Zapping of America: Microwaves, Their Deadly Risk and the Cover-up,* was published in 1977, and his second, *Currents of Death: Power Lines, Computer Terminals, and the Attempt to Cover Up Their Threat to Your Health,* was published in 1989. In the latter, Brodeur attempts to create an account of a widespread epidemic of diseases among people likely to have been exposed to electromagnetic fields. The significance of frequency-dependent differences between the ELF fields generated by power lines and the much higher-frequency RF fields[2] appears to be lost on Brodeur, since he casually mixes discussions of ELF and RF throughout the book, implying that data collected in ELF power line studies is equally valid when considering RF sources.

Brodeur has also written several books on the asbestos controversy, in which he focuses on the prominent legal actions driving the abatement industry, as well as a book on the claims for restitution filed by native New England Indians. Combined with his clear prose and gripping style, his chosen fields of investigation have led to extended commercial success for Brodeur.

Becker

Dr. Robert Becker,[3] a prominent figure in Brodeur's accounts of research into the ELF fields generated by power lines and into the U.S. Navy's proposed ELF communication system for submarines, has himself sought to popularize the discussion of the potential health effects of ELF and RF fields. Although his medical background gives him a much more credible voice in the debate, his writing, too, is somewhat more alarmist than appears warranted by the scientific data.

In *The Body Electric: Electromagnetism and the Foundation of Life,* a book he co-authored in 1985, Becker explores the function of the body's natural electromagnetic fields, linking disruption of these fields by exposure to external sources such as microwave ovens and radios to a wide range of ailments, including AIDS, heart disease, nerve damage, cancer, and emotional disorders. He explores this thesis further in *Cross Currents: The Promise of Electromedicine; The Perils of Electropollution,* discussing alternative medical practices

---

[2]See Chap. 1 for a discussion of power line frequencies and their difference from *radio* frequencies.

[3]Robert Otto Becker, born 1923, holds an M.D. from New York University.

and their possible stimulation of the body's electrical system, as well as the increase in EMF and the research into its biological effects. He does admit, however, that a primary or even cooperative causative role of electromagnetic fields is purely speculative.

Becker discounts entirely the many safety standards promulgated by various bodies around the world, claiming, "we really do not know what the safe level of RF field strength is for continuous exposure." He gives little credence to the multidisciplinary, consensus approach taken by the standards-setting bodies and their careful conservatism in translating empirical findings into standards for human beings.

Marino

An associate of Dr. Becker's while at the Veteran's Administration Hospital in Syracuse, New York (until forced out as a result of his and Becker's work on power lines, according to Brodeur), Dr. Andrew Marino[4] is now at the Louisiana State University Medical Center in Shreveport, Louisiana, where Becker is a Clinical Professor of Orthopedics. Like his mentor, Marino's credentials give his input to the RF debate a weight that Brodeur's lacks, but his contentions of adverse health effects from RF exposures under the limits of the prevailing standard (PS) suffer the same flaws. While the greater part of his research and experience relates strictly to ELF fields, this has not prevented Marino from participating in the RF exposure debate. He is often called as an expert witness in legal proceedings and is cited by groups opposed to the construction or modification of broadcasting and communications facilities. Unfortunately for those who sincerely believed Marino and who had relied upon his expert judgment in these matters, Marino himself may not attach the same credibility to that expertise; it is reported that, when asked in a land-valuation case involving power lines whether he was testifying as an expert, he answered, "I am speaking as a human being."

Popular press coverage

As public awareness of wireless communications has grown with the advent of cellular telephones and pagers, so has general awareness of the infrastructure required to make these systems available. People naturally have questions about them, and the media, sensing a market, attempt to provide answers. The press prospers by an impression of urgency, of course, so the alarming story, especially with personal trauma, is always preferred over an analysis of research.

---

[4]Andrew Anthony Marino, born 1941, holds a Ph.D. from Syracuse University.

This same phenomenon occurs at public hearings, where decision makers must hear claims from both sides of the debate ("RF is dangerous" versus "The evidence points to its safety"). Unfortunately, a review of the studies by ANSI or NCRP or EPA does not seem to provide an adequate antidote to the personal anecdote: "Two of my neighbors in the last year have died of cancer." The difficulty is in educating commissioners, council members, and supervisors on such issues as correlation not implying causation and as how truly tiny are the field strengths from many RF installations. However, the popular press does not encourage objective inquiry or understanding.

One example of the difficulties encountered when the popular press reports on scientific studies can be seen in press accounts of a report released by two researchers in 1962. This early experiment involved 4½-min exposures of mice to 9 GHz radiation at 100 mW/cm^2, a dose massive enough to raise the core temperature of the mice by 3.3°C in that short time. This went on 5 days a week *for over a year* and, remarkably, the researchers found little to report on. They did report an increased incidence of "leukosis," which they defined as a "noncirculatory neoplasia of white blood cells" but which was misinterpreted and misreported subsequently as "leukemia." Moreover, several later researchers have found deficiencies in the work and, especially, in its reporting; no criteria were established to define "leukosis," and no records of the occurrence of this condition in control groups were provided. In fact, it was later learned that there had been an endemic infection in the colony of subject mice, from which many died, and the reported data even suggested that those mice not in the control group actually had enhanced survival. Nevertheless, on the basis of such early work, not subsequently corroborated by other researchers, one still hears the assertion, "RF causes leukemia," not always even followed by "in mice."

Another example of the ways in which the media help to fuel widespread concern can be seen in the coverage of a Florida woman's lawsuit claiming that her use of a hand-held cellular phone caused a brain tumor that eventually caused her death (discussed in more detail later). This case, and the completely unrelated deaths of two high-level executives from brain tumors at about the same time, struck a nerve in the press and the public alike. Stories about the "new risks" of hand-held cellular phones appeared in nightly news broadcasts, newspapers, weekly newsmagazines, and specialty magazines including *Fortune, Business Week, The Economist,* and *Popular Mechanics.* Although many of these reports attempted to strike a balance between the charges alleged in the lawsuit and the technical basis for such claims, the overwhelming impression left with the casual observer was of the issue "cancer" being discussed in the same breath as "cellular phones." The sales of hand-held cellular phones dropped sharply soon after this media attention, but they rose again as people realized that fears of an epidemic of brain tumors were unwarranted; the pace of sales today is torrid.

Equally egregious examples of the spread of misinformation can be found in local press coverage. Stories on multiuser broadcast sites, new cellular base stations, and ordinances imposing limits on antenna construction often cite "health risks of electromagnetic fields" as primary motivations for new restrictions. It is such matter-of-fact statements that perpetuate public perceptions of risk and create the impression that antenna owners and government bureaucrats display complete disregard for public safety. Rather than taking the opportunity to further awareness and understanding of these complex issues, the media simply report as fact something considered "common knowledge."

The issues that these examples raise point to fundamental impediments to the core mission of the media. Since all media exist in a competitive marketplace, battling each other for the eye, ear, and mind of the consumer, substantial efforts are made to grab viewers, listeners, and readers and to hold their attention for as long as possible. Attention-grabbing "teases" and "soundbites," all carefully crafted examples of hyperbole, are employed by most media outlets in pursuit of this goal. While the use of such devices probably improves a station's ratings or increases a paper's circulation, it has a deleterious effect on the quality of the information they contain, often oversimplifying the issues addressed and leading to misperception and misunderstanding by the consumer.

An exception to the typical media coverage of concerns over electromagnetic field safety is an episode of the 1996 PBS *Frontline* documentary entitled "Currents of Fear." This program presented a balanced view of the key issues in the debate, juxtaposing the concerns raised by Brodeur, Becker, and Marino with the views of mainstream researchers in the field. It also presented with clarity the fact that the few scientific studies purporting to establish a link between electromagnetic field exposures and adverse health effects in human beings were found to be inadequate for statistical or methodological reasons, including those often cited by Brodeur and his colleagues.

Media's role

Despite the drawbacks identified above, the press fulfills, almost by definition, its "role" in this issue just as well as in others (*i.e.*, providing investigation, leading to education). While one might complain about the *balance* of reporting, especially from particular writers or in particular publications, the whole of the journalistic effort cannot help but be positive. People today, whether as consumers, residents, workers, or administrators, are better educated on the topic of RF radiation, and thus they are able to make more informed decisions when buying equipment, choosing where to live, taking occupational safety precautions, and recommending agency action.

Deep Pocket Risks

In the litigious haven that contemporary American society seems to have become, there are innumerable plaintiffs' attorneys anxious to plead cases against large companies in the hope of securing large settlements or even larger judgments. This has led, unfortunately, to a few suits alleging injury from exposure to RF radiation.

Tobacco is the substance most often remarked upon that was once considered benign and is now known to be otherwise, and the utility of that argument is not lost on opponents of new or expanding RF systems. As discussed in Chap. 1, however, the argument does not hold up under scrutiny. Decades of research into tobacco smoking showed clear correlation between that practice and adverse health effects, while the decades of research into RF exposure show *no* correlation with adverse health effects at exposure levels meeting any of the standards. Just as there are still individuals who argue that smoking is not dangerous, there are also individuals who still argue, without supporting scientific data, that a danger from low levels of RF must nevertheless exist.

The two brain cancer cases discussed later represent a fundamental shift in relationships and may foretell larger battles ahead. In the past, the "deep pockets" with regard to RF were admittedly big companies, but only the biggest of them was the size of the two tobacco giants, for instance, as shown in Fig. 4.1. Moreover, only a small percentage of the public at large was affected by RF emitted from the operations of these broadcasters, since most transmitting facilities were located in areas of low public use.

Company	*Net Worth (in millions)*
GE (NBC)	$ 25,824
Capital Cities/ABC	3,572
Chris Craft	1,258
Westinghouse (CBS)	1,045
Belo (A.H.) Corp.	346
Philip Morris	11,627
RJR Nabisco	9,070

Figure 4.1 1993 "giants."

Company	Net Worth (in millions)
GE (NBC)	$ 30,700
AT&T	20,000
Bell Atlantic	19,100*
SBC	14,700*
Disney/Cap Cities/ABC	14,400
Motorola	13,150
MCI	10,100
Ericsson	6,075
Sprint	5,155
Airtouch	3,860
Philip Morris	13,840
RJR Nabisco	10,065

**Size following pending mergers*
All figures projected by Value Line

Figure 4.2 1996 "giants."

Now, with the explosion in wireless personal communications, both factors have changed. First, the percentage of the population affected probably already exceeds 50 percent, since cellular telephones seem to poke out of every other purse or briefcase; RF transmitters are now local (and hand-carried), and "second-hand RF" is growing. Second, there is now a new set of companies transmitting RF. They, or their parents, are now huge, as shown in Fig. 4.2; seven of them are already larger than the second largest of the cigarette companies.

The litigation risk implied by these trends is large and is growing. In fact, this is partly the reason for the funding contributed by the industry toward further research (see Chap. 5). Unfortunately, even if the research findings continue to be negative, the threat of an adverse judgment is a powerful inducement for the big companies to settle. While they would certainly do so with stipulations as to secrecy, the small chance of a windfall can be expected to draw the plaintiffs' attorneys.

Unsubstantiated Claims

Legal cases

Not many legal cases in which RF radiation was alleged to have harmed human beings have been tried. It is assumed that some suits must have been settled quietly; such settlements almost always contain nondisclosure provisions so that they cannot be used to set precedent. There are only two cases that legal databases show as having been filed in federal court; these are discussed below, followed by a third in progress.

Foster, Forsling, & Jessop v. United States and RCA. The first case related to an incident of exposure to high fields suffered by workers on an Air Force radar installation in Alaska. In 1994, the plaintiffs sued the United States and RCA, manufacturer of the control circuitry, for failure to ensure that proper safety precautions were installed and followed. Apparently, during maintenance, all four of the safety interlocks were disabled by the work crew, and another member of the crew turned on the facility while the workers were inside the radome surrounding the radar antenna. The courts dismissed the claims against the United States on the basis that the actions of the Air Force involved "judgment susceptible to policy analysis" and therefore was not actionable. The claims against RCA were also dismissed on the basis that the "extraordinary series of mistakes" by the man at the controls constituted the superseding cause of the accident, not anything that RCA had failed to do. This case will have little effect as a precedent, since the issue of health effects due to RF radiation was not heard.

Reynard v. NEC and GTE. The second case is the more (in)famous one. It alleges that the brain tumor from which Mrs. Susan Elen Reynard died in 1993 in Florida had been initiated, or else its growth had been accelerated, by the woman's use of a cellular phone. She is reported to have begun use of the phone in August 1988. An MRI in March 1989 showed that a tumor was present above her left ear. A second MRI in May 1990 showed that it had increased in size. She underwent surgery in 1991 and survived for 2 more years.

Apparently, a suit was filed by the woman and her husband against the phone manufacturer and the cellular operator in 1992. That suit was dismissed without prejudice[5] in 1993. In 1994, a second suit was filed, to which the defendants filed a motion for summary judgment.[6] Arguments on that motion were heard in 1995 in U.S. District Court in Tampa, and the motions for summary judgment were granted in May 1995, meaning that the plaintiffs had failed to "introduce evidence which affords a reasonable basis for concluding that it is more likely than not that the conduct of the defendant was a substantial factor in bringing about the result."

This case is important because arguments were heard in regard to the possibility that RF radiation had either initiated the tumor or ac-

[5]This term means that no judgment was rendered and the plaintiff is allowed to file the same or a similar action, at his or her discretion.

[6]This is equivalent to a dismissal *with* prejudice, which means that the plaintiff would not be allowed to file again.

celerated its growth. The statements submitted by the plaintiffs' experts were less than conclusive. One of the experts cited is Dr. John Holt, who is quoted as believing that human cancer cells die when exposed to RF radiation at 434 MHz, at high power, and that those exposed to the same radiation at low power exhibit a phenomenon he calls florescence. This belief comes from his experimental work with 434 MHz radiation as a treatment mode in conjunction with ionizing radiation. In response to the observation that the cellular phone operates between 825 and 845 MHz but not at 434 MHz, Dr. Holt stated, "I don't know the answer," as to the effects of the cellular phone's frequency. He declined to "hazard a guess" because "everything is so individual in the cancer field."

David Perlmutter, M.D., was the other of the plaintiffs' experts, and the court found his statements, though strongly worded, to be speculative:

> It is certainly quite evident that both animal studies and human brain tumor cell studies provide strong inferential data that the use of this device represents a clear health hazard and would likely accelerate the growth of brain tumors in humans.
>
> It is my opinion, within a reasonable degree of medical certainty, that the use of the hand-held cellular telephone did in fact accelerate the growth of Susan Reynard's brain tumor.

That finding was made in connection with Dr. Perlmutter's further statements that he was "in complete agreement" with the defendants' expert opinion that:

1. The tumor had begun to grow many years before she began using the cellular phone.
2. "No scientific or medical studies have shown that exposure to emissions deposited at the brain from a source such as a portable cellular telephone operating at a power level and frequency of the portable cellular phone alleged to have been used...is associated with adverse biological effects, including initiation of brain cancer or promotion of brain cancer growth."

Wright v. Cellular Telephone Industry Association, *et al.* Filed in 1995 in Cook County, Illinois, by a woman who has developed a brain tumor, the suit alleges 1) that data exists showing the harmful effects on human health of cellular telephone use and 2) that a conspiracy exists to suppress that data. While several defendants have already been dismissed from the action under the second allegation, the first allegation may yet result in health arguments being heard.

Property values

Diminution of residential property values is one of the common claims used to argue against installation of new wireless communication towers. The importance of this argument is the implied concession that there is probably nothing to the health hazard claim, at least nothing that the opponent expects to prevail. For these people, FDR's famous line "We have nothing to fear but fear itself"[7] may be true. The argument is made that the mere *fear* of a health hazard is sufficient to be an actionable cause. Since properties are so unique, and since the residential property market is so illiquid, such claims are nearly impossible either to prove or to disprove.

[7]President Franklin Delano Roosevelt (1882–1945) was seeking to encourage the American populace that the Depression could be survived.

Chapter

5

Future Concerns

Further Research

Research into the biological effects of RF radiation began over 100 years ago, with its first use in diathermy. Modern research has been actively conducted since the 1950s and continues today at a greater pace than ever. The gross interactions of mammals with RF radiation have been fairly well quantified in terms of SAR, and most research now focuses on two major areas of inquiry: athermal effects[1] and interference to biomechanical devices. The latter devices are, in particular, electronic hearing aids and implanted pacemakers. Since interference to such devices is not a biological interaction *per se,* it falls outside the scope of this text.

Athermal effects are those reportedly observed at power densities and SARs so low that localized tissue heating does not occur and core body temperature does not rise. The suggestion has been made in several quarters, for instance, that RF may be a cofactor in cancer initiation or growth. By itself, in other words, the RF would not be adequate to generate the biological endpoint of tumor formation, but in the presence of other environmental factors it may *contribute* to the formation of a tumor.

Research findings have thus far not been supportive of this suggestion, at least in the combined judgment of the experts on the NCRP and IEEE panels actively studying those findings. What has been demonstrated is that certain effects can be observed in the laboratory with *in vitro* studies. Most of this work has required the use of pulsed

[1]*Athermal* means, literally, without heat(ing).

or heavily modulated RF at pulse or modulation rates in the ELF region. If athermal effects of RF are found to have adverse biological reactions in human beings, the data suggests that those effects will occur only within certain ELF modulation "windows."

WTR

The best publicized research effort is being conducted under the aegis of the Wireless Technology Research, L.L.C. (**WTR**). Here is the group's own description of its founding:

> In February 1993, the cellular telecommunications industry made a public commitment to support independent and rigorous scientific research into the safety of portable cellular telephones and other wireless communications technology. The Scientific Advisory Group (SAG) on Cellular Telephone Research was subsequently established with criteria and procedures guaranteeing absolute non-interference by industry to assess the potential public health impact of wireless technology and to recommend corrective interventions where necessary.

Some observers think that WTR "doth protest too much," judging that its ongoing public relations campaign is a significant part of its mission (*i.e.,* to provide cover for the cellular industry). The founding of the SAG did, indeed, deflect much scrutiny at a critical time from the cellular industry itself. In light of the relative paucity of data on localized exposure at ~850 MHz, and in light of the highly publicized death of Susan Reynard in Florida, the industry was not in a strong position to respond to allegations of health problems.

Nevertheless, the 5-year, $25-million research program has been active and has commissioned appropriate studies. The most encouraging aspect of the WTR enterprise is the important position held by Dr. Arthur W. Guy. This respected researcher chaired Subcommittee IV that developed the ANSI-74 revisions, he chaired Subcommittee IV that developed the major ANSI-82 revisions, he chaired NCRP Scientific Committee 53 that developed NCRP-86, and he was Vice-Chair of IEEE SCC-28 when IEEE-91 was developed. His continued, direct involvement with WTR provides reassurance that solid research is to be conducted.

In "Phase One: Laying the Foundation," the SAG organized itself into various Surveillance, Research, Risk Management, Product Labeling, and Outreach groups, adopted Good Laboratory Practices, Good Clinical Practices, and Good Epidemiology Practices, set forth four Guiding Principles, produced a beautiful 32-page brochure, and spent $3 million. In 1995, the SAG added an administrative layer, Wireless Technology Research, L.L.C., "to ensure the proper structural and financial support for the research program." That year also saw the commencement of "Phase Two: Extramural Research," on which WTR expected to spend more than $10 million. The concluding Phases were to be:

Phase Three: Short-Term Study Findings (1996)

Phase Four: Long-Term Study Findings (1997)

Phase Five: Informed Judgments (1998)

The funding for WTR is provided by a voluntary annual "health and safety assessment" paid by cellular equipment manufacturers and service providers, virtually of all them members of the Cellular Telecommunications Industry Association (**CTIA**). The CTIA, under the leadership of its president, Tom Wheeler, has since expanded the scope of WTR's inquiries to include pacemaker interference, hearing aid interference, and driver safety. Certain manufacturers have subsequently withheld payment on their 1996 assessments because of concerns about the loss of focus of the WTR program and about the expensive administration costs.

Another issue is the suit filed in Cook County, Illinois, that named WTR Chairman Dr. George L. Carlo and his own firm, Health and Environmental Sciences Group, Ltd., as defendants. The suit discussed in Chap. 4, alleges that the plaintiff developed brain cancer as a result of her cellular telephone use and that the defendants, including CTIA, had conspired to conceal from the public information about possible harmful effects of cellular phone use. Dr. Carlo and his firm were dismissed from the suit in December 1995, shortly after WTR was added to the complaint, amid questions about use of WTR funds for Dr. Carlo's legal fees.

Of more import, however, are questions that have arisen again regarding WTR's true independence from industry influence. It is reported that CTIA has directed WTR to halt research on dosimetry certification on the rationale that it duplicated similar work being undertaken by several member manufacturers. This had renewed suggestions that the whole SAG/WTR effort was undertaken as public relations work, to defuse the concern about the Florida case and to forestall any governmental regulatory efforts. Combined with a continued funding shortage, the possibility exists that WTR may not complete its ambitious undertaking.

The one study funded by WTR that has been published appeared in the May 1996 issue of *Epidemiology*. The study, conducted by Epidemiology Resources, Inc., in Boston, compared mortality rates among over 250,000 cellular phone customers who had used their phones for at least 3 years. Reportedly, those using the portable style of phone, with a built-in antenna, actually had a reduced mortality rate compared with those whose phones were installed in vehicles and presumably had the antenna mounted at an outboard location. The study has not yet been reviewed by other researchers and the significance of the findings has yet to be determined, but it was, perhaps not surprisingly, well received within the industry.

Other research

Unfortunately, WTR was so well funded and organized that most of the researchers in the field of RF biological effects have developed a relationship with that group. Whether as members of one or more of its groups or as researchers under contract, most have followed the WTR plan of research, and it will take some time for them to seek new funding and to develop new plans should WTR not continue its work. Even if it does continue, it appears to be almost 2 years behind its original schedule for allowing "informed judgments" to be made on the carcinogenicity of cellular telephone use.

New Equipment

Induced current meter

Until 1996, it was possible to measure body currents reliably in only a limited number of settings. As discussed in Chap. 3, both Narda and Holaday had introduced **parallel-plate body current meters,**[2] but these meters, shown in Fig. 3.4, exhibited severe calibration problems, including negative readings, in complex fields involving more than one frequency. When the RF exposure assessment was being done in a limited area with a single frequency, such as might be the case with heat sealers or other ISM equipment, the parallel-plate meters could be used for measurement of body current. In other settings, however, such as broadcast sites, reliable equipment for measuring body current was not commercially available. This meant that compliance with the new body current limitations introduced by NCRP-86 and ANSI-92 could not be determined at that time.

To address this problem, the author's firm did develop in 1993 a parallel-plate boot, as shown in Fig. 5.1, and calibrated it for use at AM transmitting sites. It was found to be quite practical for taking

[2]Often called "bathroom scale" meters, due to their obvious resemblance.

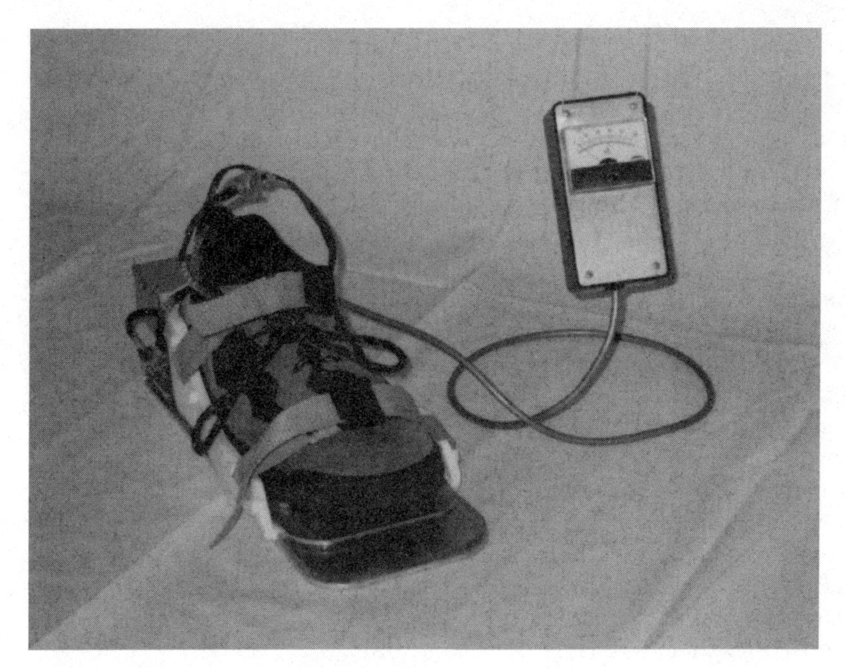

Custom design for measuring induced current at AM frequencies.

Figure 5.1 Parallel-plate boot.

the many measurements at a site needed to establish compliance with ANSI-92. By 1996, however, Holaday had introduced a reliable **toroidal anklet meter,** first proposed by the author in 1993, which coupled much better to the current flowing in a person's leg and provided a reliable, accurate means of measuring body current (see Fig. 3.5). As a result, RF exposure surveys since 1996 should include assessment of induced body currents.

For the first time, the *engineer* making the RF exposure survey becomes *part* of the measurements. While mental decisions made by the engineer previously affected the nature of the measurements, now the physical composition and condition of the engineer also affect the results of the measurements, as discussed in Chap. 2. The hallmark of scientific inquiry is repeatability, but the use of the engineer taking the measurements as part of the measurement apparatus introduces a significant, uncalibrated variable. Therefore, some standardization of measurement protocol for induced body current measurements must be adopted in order for the measurements to have meaning. Especially as a safety issue, standardization is important. In 1993, the author urged[3] the FCC to set forth, as part of its then-imminent adoption of ANSI-92, simple measurement protocols to ensure consistency in findings. The FCC's adoption of a new standard has become highly politicized, and it is still not clear whether improved measurement standardization protocols will be included.

There are four elements to a proper standardization program for measuring induced body current. The text of ANSI-92 specifies the limit on induced body current, "as measured through each foot," and it is intuitively clear (as well as having been demonstrated) that the current through a foot will be maximum when the other foot is lifted. Since movement about an area subject to high RF fields generally requires walking, one foot is lifted much of the time. Even in a fixed location such as an operator position at an RF heat sealer, the operation may require stepping to the side and back again for activation or for passage of the material. Especially in light of ANSI-92's 1-second averaging time for body current, it is reasonable, as well as conservative, to conduct induced body current measurements *with one foot raised* (element 1).

Due to the considerable variation in isolation from ground provided by different shoe materials, sock materials, and sock moisture levels, it is recommended that engineers taking the measurements wear a *metal plate on the bottom of the shoe* (element 2) on the leg to which the current meter is attached. This will ensure that the shoe material variable is removed from the measurements. It is also recommended that the plate have a braided metal strap attached near the heel and that this *strap be wrapped around the engineer's bare ankle* (element 3); this removes foot moisture as a variable.

It has also been demonstrated that the induced body current is significantly affected by the height of the engineer making the measurements (and, to a lesser extent, by the engineer's girth). Particularly at AM radio frequencies, the taller the engineer, the higher the induced current. It is recommended that the engineer *raise one hand to a height of 2 m* (element 4), approximating a condition met by more than 99 percent of the population (body height not exceeding 6 ft 7 inches). A length of rope or nylon strap can be attached to the engineer's belt or to the foot plate; with a loop at the other end for his or her hand and with the length adjusted properly, it becomes quite easy for the engineer to raise his or her hand to 2 m above ground with fair precision and high repeatability.

---

[3]Technical presentation at the National Association of Broadcasters Engineering Conference, April 21, 1993, plus comments filed January 21, 1994, in the Docket 93-62 proceeding.

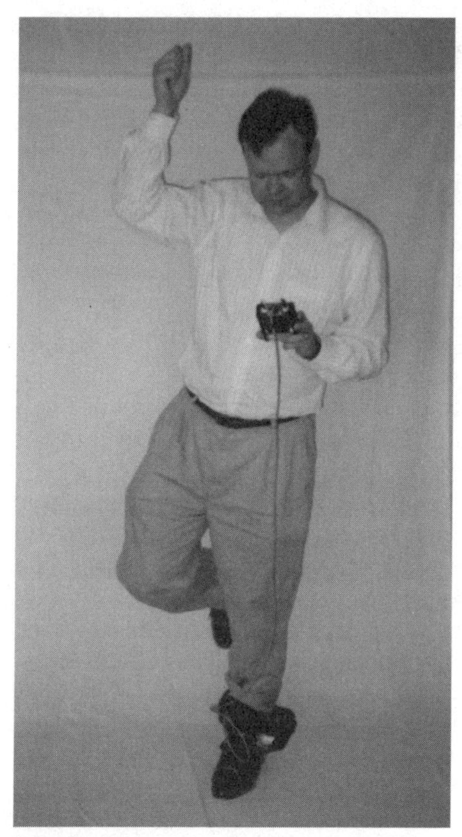

Note all four elements: one foot raised, metal foot plate, ankle strap, and hand at 2 meters.

Figure 5.2 Engineer "in position."

Figure 5.2 shows an engineer performing induced body current measurements with the proper equipment (foot plate and strap) and with the proper stance (the other foot lifted and one hand raised). With all four standardization elements included, the conditions of induced body current flowing to ground are standardized, and the engineer can be ensured of observing worst-case conditions at each measurement location.

Contact current meter

Using the ground-interface standardization protocols for induced currents, it would seem an easy matter to perform contact current measurements simply by touching metal objects about the site and observing the current flow with the anklet meter. However, this would raise serious safety concerns for the engineer conducting the measurements, since a good electrical path would be presented to voltages on metal objects touched, and contact currents through the engineer's body could be high. In recognition of this, ANSI-92 limits only "RF current through an impedance equivalent to that of the human body for conditions of grasping contact as measured with a contact current meter."

Therefore, the proper approach is to take the engineer *out of the circuit* for assessing compliance with the contact current guidelines. Unfortunately, as of late 1996, reliable, calibrated meters still were not commercially available. Moreover, ANSI-92 itself only defines the human impedance up to 3 MHz, while the Standard requires compliance at frequencies up to 100 MHz. Figure 5.3 shows the data provided by ANSI-92. This creates a situation in which it is simply not possible[4] to determine compliance, since the impedance to be used in the required "contact current meter" is not defined.

The Canadian Safety Code 6 contact current limit extends only to 30 MHz, and that document sensibly provides information on assumed human impedance over its full range. One would expect that Code 6 could provide some guidance for attempting to comply with ANSI-92, at least for frequencies between 3 and 30 MHz. However, the impedance data from Code 6 does not agree well with ANSI-92. At 3 MHz, for example, ANSI-92 gives a human impedance of 400 $\pm$ 90 ohms, while Code 6 gives 900 $\pm$90 ohms, more than double. Agreement at 10 kHz, the lowest frequency for which ANSI-92 gives data[5] and the bottom of the range for Code 6, is even worse: ANSI-92 says 580 $\pm$120 ohms, while Code 6 says 1550 $\pm$150 ohms, almost triple. Apparently, this is due to the development of the Canadian data from measurements in which the subject *touches* the current-conducting object, while the ANSI data is based upon a *grasping* contact. Of particular concern is that the impedance is trending *down* as the frequency goes up. This is a critical factor in the attempt to extrapolate the data for use up to 100 MHz, since a lower impedance would give a higher current, making it more likely for a given situation that the guidelines are being exceeded.

Actually, the situation is not as bad as it would appear from ANSI-

[4]Insertion of the human subject back into the circuit would make measurements up to 100 MHz possible. However, such measurements would be uncalibrated and could not ensure that a worst-case assessment was being conducted.

[5]Unfortunately, the human impedance data given by ANSI-92 does not reach to the *bottom* of its required frequency range, either. Figure A6 in ANSI-92 stops at 10 kHz, while the guidelines are specified, at ever tighter limits, down to 3 kHz.

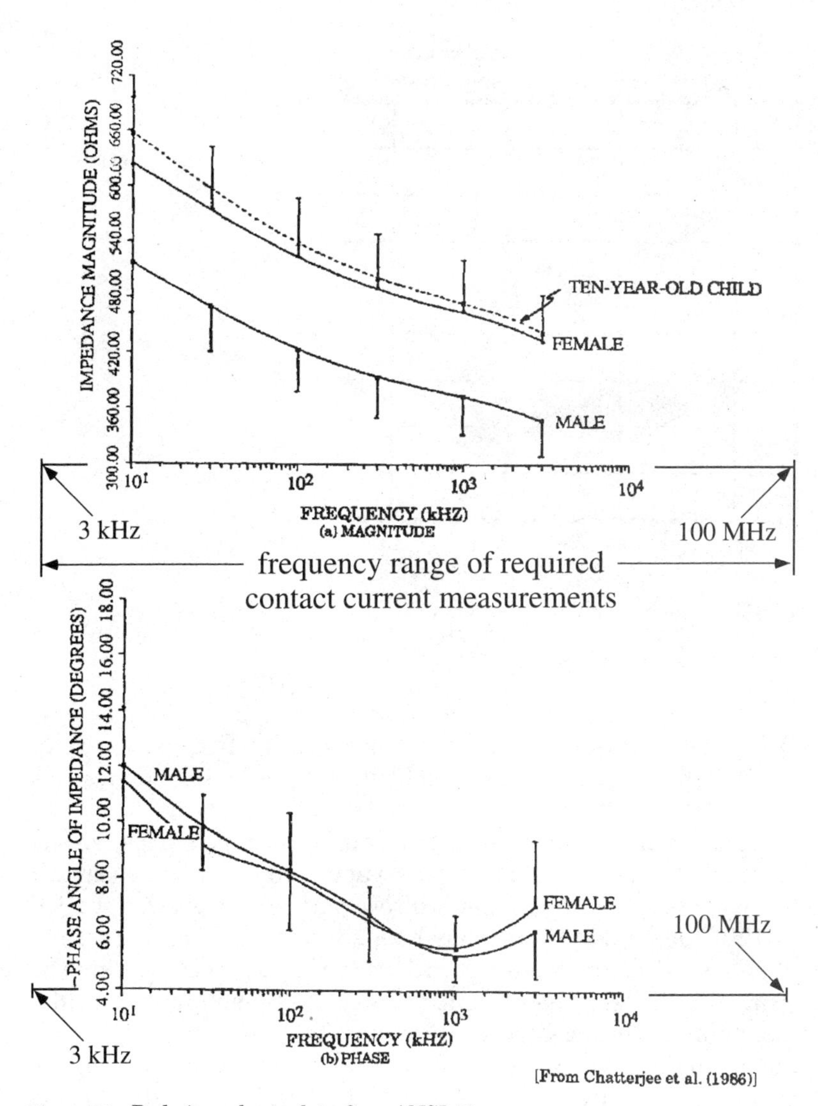

Figure 5.3 Body impedance data from ANSI-92.

92's Figure A6, which has the abscissa crossing the ordinate axis at an arbitrary level (300 ohms), not at zero. When plotted correctly, as shown in Fig. 5.4, there is a more realistic trend apparent. The only disturbing factor is the drop in the Code 6 impedance beyond the range of data included in ANSI-92. In order to perform full ANSI-92 RF exposure assessments, *some* impedance level through 100 MHz

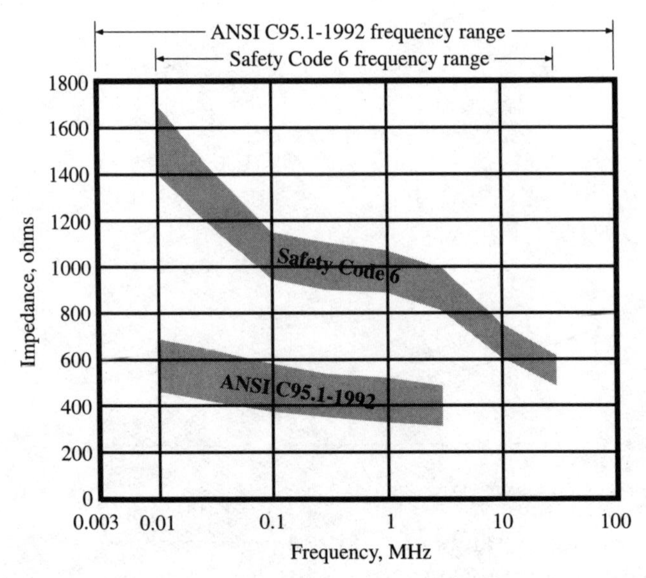

Figure 5.4 Human equivalent impedance data.

must be assumed. Therefore, the author has conducted independent research[6] to collect actual human impedance data at frequencies up to 100 MHz, which has been used to establish worst-case impedance values.

Aside from the matter of selecting appropriate impedances, an actual metering apparatus can be constructed in a straightforward fashion. The concept is to provide a direct path to ground for the RF (or ac at 120, 208, or 240 V that may be inadvertently exposed) from a touched object, with that path not including the engineer making the measurements. Routing the current through the same anklet toroidal meter used for the induced current measurements provides the calibrated metering necessary. As shown in the schematic of Fig. 5.5, use of a "fixed" ground contact (the substitute foot) and a "movable" object contact (the substitute hand) provides flexibility in use, and the addition of the human equivalent impedance completes the apparatus. Figure 5.6 is a picture of the finished prototype.

An important consideration when using contact current meters is ANSI's 1-second time averaging requirement. This is based upon the

[6]This work is awaiting publication.

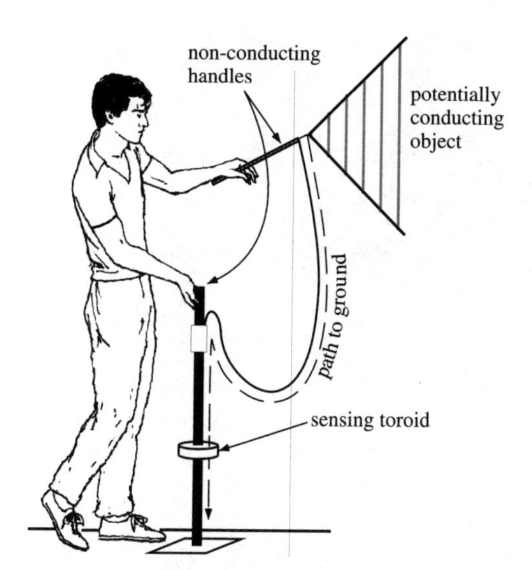

Figure 5.5 Basis of custom contact current meter.

presumption that the current flow will come from an electrically isolated object, such as a car on rubber tires, that is immersed in an EMF and has become capacitively charged. Therefore, the "peak hold" feature of the contact current meter needs to be used in order to ensure that the highest, instantaneous current flow is measured. A condition of lower, steady-state current flow may also exist, but the measurement procedures need to be designed to assess worst-case conditions.

Information Superhighway

It is commonly acknowledged that we are in the midst of a technological revolution. Even in the early 1990s, it was unusual to find a personal computer in most homes, and cellular telephones were normally hard-wired into the cars from which they operated. By 1996, though, it was a surprise when someone did *not* have a PC, Mac, or clone at home, and cellular phones were now more often battery-powered and carried in briefcases and purses. Both trends foretell a rapid advance of technology in the next decade, and in many cases that will mean an increase in ambient and specific RF levels in many locales.

Personal communications—hand held

There are two, and only two, cellular companies licensed in each market. The FCC has licensed one to a wireline common carrier, such as

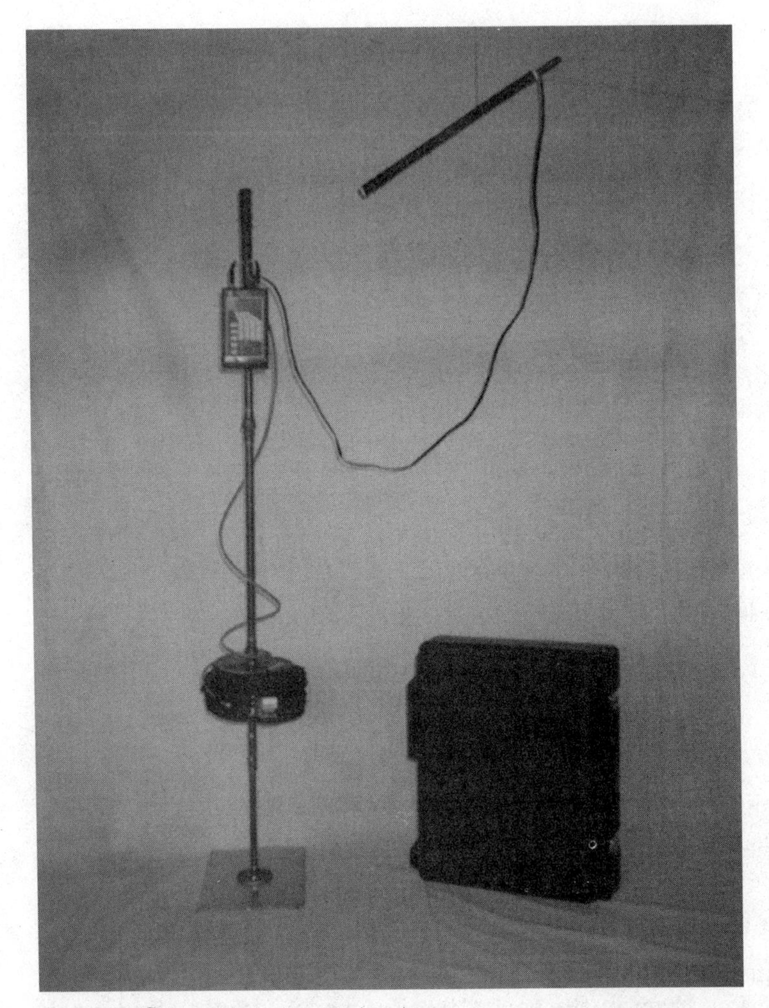

Figure 5.6 Custom contact current meter.

GTE or **AT&T,** and the other to a nonwireline competitor, often affiliated with **CellularOne.** The rapidly growing popularity of this service was sure to lead to more competitors, almost regardless of FCC action, and there developed, under the names **Nextel** and others, a service using the **Specialized Mobile Radio (SMR)** frequencies. These were packaged in various markets as a combined dispatch and mobile communications service, with some success.

The FCC, however, opened up the mobile communications market further in 1995 and 1996 with its auction of frequency blocks for **Personal Communications Services (PCS).** It is reported that in

Japan, where PCS technology is already several years old, employees will let the company phones on their desks ring ("must be an impersonal call") while answering their pocket phones right away ("must be important, whether business or personal, if they have that number"). Although the frequencies for PCS are higher (about 1.9 GHz versus the 800 to 900 MHz used by cellular and SMR), the service and system configurations are similar. This reassigned spectrum will create room for about seven or eight competitors, which the major markets will certainly accommodate. Lesser markets may not see as many systems built right away (or perhaps ever).

To support these new systems, there will be a proliferation of RF-emitting base stations. These are generally low-power units, mounted on existing structures and spaced much closer together than the cellular base stations. Interconnection is done by various means, including hard-wire telephone, cable television backbone, and occasionally microwave. Generally, there are so many sites required that microwave interconnection cannot be widely accommodated due to spectrum limitations and cost, although microwave may still be used for remote sites. One type of installation is shown in Fig. 5.7.

From the standpoint of RF exposure, the multitude of new sites should have little effect on most exposure situations. The low power of these small base stations will ensure that only a very close approach, say, within 3 ft, could involve exposure conditions near the PS guidelines. Since these stations will often be mounted on poles or overhead wires 20 to 40 ft above ground, nonoccupational exposure situations would not arise. When mounted on rooftops or building

Figure 5.7 Overhead base station.

faces, the same concerns arise as are involved with cellular antennas now, although the lower power levels will make compliance easier to achieve and mitigation, where needed, easier to implement.

Not only is the number of service providers growing, so too are the services that they offer. PCS may seem like cellular at first, but it is just the beginning of a wide range of products. Paging, messaging, news alerts—all these will proliferate as the spectrum becomes available and the supporting infrastructure is built.

Some prognosticators suggest that every American will eventually (soon?) have his or her own personal telephone number. Of course, each of us already has a personal number: our social security number. One acquires that number right after birth, with tax deductibility as the incentive to ensure that everyone does so, and it just happens, of course, that there are 10 digits in both phone and social security numbers. With a billion of each, the match could indeed be made.

With more efficiency than even Winston Smith[7] feared, the "phone" systems around the country must track individuals' locations in order to route their "calls" through. Thus, turning on your personal communicator would create a record of your location and the time at which you were at that location. Records of this type are *already* created for those persons who use cellular phones, but concurrent advances in computing power and data storage technologies make Big Brother an increasingly realistic prospect.

Personal communications—fixed

As with many categories within wireless communications, the one used here, hand held versus fixed, is becoming more blurred. Personal computers have traditionally been connected to distant computers or networks via **modems,**[8] which work over wired telephone lines. There are now wireless data links using infrared frequencies that enable limited, local portability for the computing device, and there is a growing array of products that use RF for **wireless interconnection.**

There are two broad categories of wireless networks in present use, distinguished principally by size and data rate: WWAN and WLAN. These terms are the familiar wide area network and local area network, with a prefix denoting wireless. WWAN covers a city or region,

[7]The protagonist in George Orwell's novel *1984*.

[8]The word comes from *mo*dulate-*dem*odulate, which is what the devices do to their digital signals in order to make efficient use of the analog telephone system.

with a relatively low data rate. It uses several RF bands, including Cellular Digital Packet Data (CDPD), which uses idle time on existing analog channels, with automatic channel hopping. Deployment of digital cellular telephone and PCS networks will support data transfer at higher data rates. WWAN also uses specialized data-only packet radio systems, operating in the 400, 800, and 900 MHz band on assigned channels; only 2 to 3 W of power is used by these systems. WWAN users include low-volume applications such as dispatch, credit card verification, traffic and weather advisories, and some two-way messaging for pages and e-mail.

WLAN, on the other hand, provides a higher data rate over a smaller service area, a single building or complex being a common application. In essence, WLAN is providing wireless access ports to an existing, wired LAN. Individual device transceivers are mounted on both fixed and portable computers, and RF power of 50 to 200 milliwatts is used at frequencies in the 900 MHz, 2.4 GHz, and 5.7 GHz bands. Spread-spectrum modulation is used for resistance to multipath and other interference, and the WLAN provides flexibility, especially in dynamic settings such as warehousing, manufacturing, and disaster recovery.

As progress continues with the use of RF spectrum for digital data links, the time may come when the response of computers to network inquiries is technically no different from the response of phones to voice communications. "Full-time" connectivity through multiplexed radio links is on the horizon.

HDTV

Despite the FCC's attempt to set the format to be used for the wireless transmission of video signals, there is a natural convergence of computer and television technologies. If local TV stations are required to use a specific transmission standard for **High Definition Television (HDTV)** or even **Advanced Television (ATV),** the result will be principally to marginalize those stations, as computer capabilities work with broadband signal delivery mechanisms to effect sophisticated, personalized information retrieval, data processing, and communications. Those broadband delivery capacities may be provided by cable, by telephone, by other segments of the RF spectrum, such as the 2.5 GHz wireless cable, or by satellite, but the specific method will become increasingly transparent to the individual user. This is unfortunate, since local TV stations occupy a considerable band of RF spectrum that is well suited for point-to-multipoint service and could be a valuable part of a broad communications network. Unless the FCC backs away from its regulation of content (*i.e.,* format), those stations will become more dependent on their local nature, which will be their distinguishing feature, with news of local events becoming their principal *raison d'être.*

Regulatory Changes

FCC

There have been three major regulatory changes at the FCC, one in 1995, one in early 1996, and one in late 1996. In 1995, the requirements were clarified and tightened for RF showings of compliance that are necessary for license renewal (see Chap. 2). In early 1996, the Telecom Act was passed by Congress and signed by President Clinton; certain of its provisions are reviewed at several locations within this text (see below). As of late 1996, the FCC has taken final action on Docket 93-62, adopting a new RF exposure standard, and is still preparing updated guidance to the telecom industry on how to comply.

The author has long argued that, since ANSI-92 exists, broadcasters and others responsible for the emission of RF radiation should abide by it, regardless of the status of the FCC's politicized adoption process. To the extent that other consulting engineers have done the same, the actual adoption by the FCC should not cause much change in the practical application of the standard.

ANSI/IEEE

Inside the front of all ANSI standards appears this statement: "CAUTION NOTICE:...The procedures of the American National Standards Institute require that action be taken to reaffirm, revise, or withdraw this standard no later than five years from the date of publication." ANSI/IEEE C95.1-1992 was published in 1992, so ANSI will presumably be taking appropriate action in 1997 with regard to that Standard.

It is the policy of the IEEE Standards Board that, "Every IEEE Standard is subjected to review at least every five years for revision or reaffirmation." Therefore, since the C95.1 promulgation is handled under the auspices of IEEE, and since its standard is IEEE C95.1-*1991,* one might expect that 1996 would be the year for IEEE to take action in that regard. However, while Subcommittee IV of IEEE Standards Coordinating Committee 28 (see Chap. 2) does meet semiannually, there has been no push for revision or reaffirmation of the standard. A request from ANSI for such action was only announced at the June 1996 meeting, and it may not be possible to complete action in time, especially if revision is contemplated. This delay is probably for the best, since only after FCC adoption of ANSI-92 will there be wide regulatory review of its provisions, perhaps uncovering new problems in addition to those apparent earlier.

However, the IEEE Standards Board policy also states that, "When a document is more than five years old and has not been reaffirmed, it is reasonable to conclude that its contents, although still of some value, do not wholly reflect the present state of the art." Thus, IEEE may inadvertently be putting the FCC in the awkward position of adopting, in 1996, a standard that is just becoming officially obsolete.

Local jurisdictions

The regulatory environment at the local city and county level has been changing, with even small cities now following the fashion of applying moratoriums on further deployment of telecommunications systems and of instituting ordinances. It seems that each ordinance is different from the next, each being the product of the planning staff in that jurisdiction and reflecting that staff's concerns, knowledge, and philosophy.

The Telecom Act will also affect the regulatory ability of local jurisdictions through its mandate for federal facilities to be made available for RF facilities. Section 704(c) of that Act reads, in part:

> AVAILABILITY OF PROPERTY—...[T]he President or his designee shall prescribe procedures by which Federal departments and agencies may make available on a fair, reasonable, and nondiscriminatory basis, property, rights-of-way, and easements under their control for the placement of new telecommunications services that are dependent, in whole or in part, upon the utilization of Federal spectrum rights for the transmission or reception of such services. These procedures may establish a presumption that requests for the use of property, rights-of-way, and easements by duly authorized providers should be granted absent unavoidable direct conflict with the department or agency's mission, or the current or planned use of the property, rights-of-way, and easements in question.

There may be some argument over the choice of words (*i.e.,* whether "new telecommunications services" means only new *types* of service or whether it includes also new transmitting *facilities* for a service that was not before available in that area[9]). If the latter, for which a good argument could be made, virtually any property owned, leased, or managed by the federal government must be made available. This would include Forest Service lookouts, the post offices in every town in the country, the 12,000 to 13,000 offices of the Department of Agriculture (some in surprisingly populous areas), perhaps even aviation beacons—virtually *any* facility owned, leased, or managed by the federal government. It is likely that the military would claim the "direct conflict" provision, but that still leaves an enormous range of potential new sites. In fact, that Section of the Act goes on to state that:

> The Commission shall provide technical support to States to encourage them to make property, rights-of-way, and easements under their jurisdiction available for such purposes.

which means that all state property may be available, as well. This regulatory change, while not widely remarked upon in the press, may well be among the most significant, particularly for the new personal communications services so much in need of new sites.

Antenna Farms

The definition of *antenna farm* is changing. The farms are moving down lower, closer to populated areas, and they are becoming less visible. The principal changes at existing, *broadcast* antenna farms relate to the anticipated addition of new transmitting antennas for the new Advanced Television service. Each existing TV station will be assigned a second channel for the still-undefined ATV use. Almost all of these will be in the UHF band and, in principle, this could mean a doubling of the number of TV antennas across the country. In practice, however, limitations such as tower loading and local environmental constraints are often forcing several stations to work together toward "master antenna" systems, with one antenna or one antenna stack handling the signals for more than one station. Other stations are taking an incremental approach to ATV, mounting light-weight

[9]The wording could even mean any facility that would be new *at the proposed site,* even if it were already in operation nearby and simply wanted to relocate to government property for some reason. In light of the wording in the Congressional versions of the Act before they were consolidated in House-Senate conference, this loose interpretation is not very supportable.

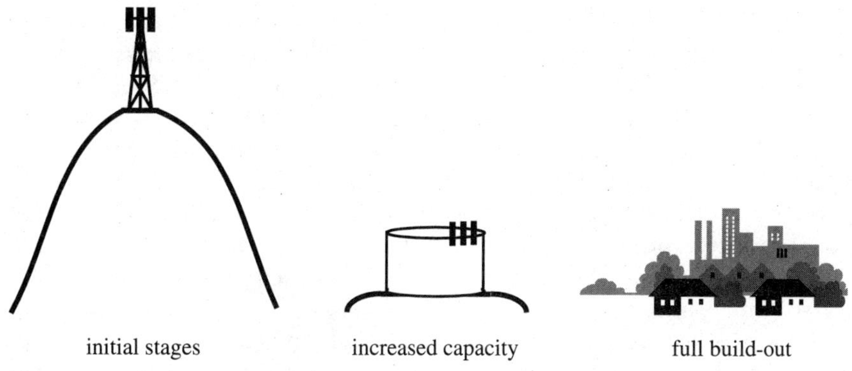

Figure 5.8 Cellular system evolution.

antennas at lower heights on their towers and waiting to see whether the demand for the new service, when it comes, will justify the expense of a more powerful installation.

It is the new, personal services, with their proliferating antenna requirements, that are driving the development of the new antenna farms. Early cellular telephone systems, the first personal communication systems, had so few customers that frequency reuse was not a significant constraint. As shown in Fig. 5.8, cell sites were selected first at the highest possible locations in order to have line of sight to as large an area as possible, and they were operated at the highest allowable powers, in order to serve those large areas. However, the FCC allotted only a certain number of channels to each provider, and once the maximum number of mobile units were communicating with the base station, that site would be at capacity and no other calls could be accommodated. Therefore, as the demand for cellular services grew, there was a general progression to adding more base stations, which could reuse the same set of available frequencies, but these base stations were sited at lower elevations and operated at lower powers, so the service cells became much smaller.

The high, hilltop cellular sites have already been abandoned in many areas; their long range would only cause interference to the network of smaller cells. Of course, the laws of physics do not change, so within each of the smaller service areas there is still a need for line-of-sight service. Sites with *relative* height are still selected, whether a downtown office building or, in residential areas, a hilltop house or a several-story public building.

With third-generation[10] personal communications systems, siting of new antennas will be done on sides of buildings, on existing light standards and utility poles, from existing overhead cables, and in other creatively selected locations from which line of sight can be obtained to an area with a significant number of potential mobile units. These areas will be, from the start, much smaller areas than the cells making up the first-generation cellular networks, since the market for third-generation equipment is believed to exist already. Demand is therefore expected to leap quickly to near-capacity levels, rather than grow slowly, as did the first two generations.

A gross estimation of the total number of new sites needed in urban areas can be made at an assumed ratio of one site per 25,000 urban population. Note that this is an especially problematic ratio, since the urban *residential* population base is the widely available measure (from the U.S. Census), but the target market is the *business* population, which can be many times the urban population due to the daytime influx of commuters. Nevertheless, one third-generation system might need on the order of 60 sites to cover Manhattan,[11] or 580 sites to cover the Los Angeles basin.[12] Even if this admittedly gross estimate is off by a significant factor, there is still an enormous need for new base station sites. Consideration should also be made of the fact that there will be several competing third-generation systems being deployed at about the same time.

For many of the systems, collocation is becoming an increasingly important consideration. In the past, collocation was considered to reduce overall site development costs, for such expenses as site lease, site improvements (roads and power), tower and/or building construction, and landscaping. Collocation today will also reduce site "development" costs, but those costs no longer relate so much to construction; a higher percentage of costs today involve professional fees to secure conditional use permits. The time saved by collocating at a competitor's existing site also represents a "cost" savings, since third-generation systems need to be almost completely deployed before the system is publicly offered. Significant coverage gaps at that time would seriously compro-

---

[10]Analog cellular systems are considered first generation, and digital cellular systems are considered second generation, with their substantial increase (up to sixfold) in system capacity. *PCS* systems, to use the FCC's term, are considered third generation, with much lower power levels and therefore smaller cells and antennas that are more closely spaced. The range of services offered by third-generation cellular systems may be broader, as well.

[11]The 1990 population of the Borough of Manhattan was 1,488,000 persons.

[12]The 1990 total population for the Los Angeles Consolidated Metropolitan Statistical Area was 14,532,000 persons.

mise the system's commercial viability. While cellular systems used to compete in terms of coverage, users now expect universal, seamless coverage, and competition is more in terms of service.

As cell sites move down and in, relative to built-up areas, the connection of cell sites to central switching facilities is increasingly accomplished by wired infrastructure, whether over dedicated telephone lines, cable television distribution and trunk systems, or dedicated copper or optical lines. The need for point-to-point microwave, which had been the most widely used method of connecting sites with the switcher, will increase only a little, if at all; in fact, as wide area sites are taken out of service, there may be a net decrease in the usage of microwave links for this purpose.

Opponents' Use of Public Fears

The ability of people opposed to deployment of technological improvements to stall telecommunications projects on the basis of RFR fears has largely been checked by the federal preemption of that issue by the Telecommunications Act of 1996. Section 704(a)(7)(B)(iv) of that Act states:

> No State or local government or an instrumentality thereof may regulate the placement, construction, modification, or operation of personal wireless services facilities on the basis of the environmental effects of radio frequency emissions, to the extent that such facilities comply with the Commission's regulations concerning such emissions.

While the preemption is nominally limited just to PCS (definitely), cellular (probably), and paging (possibly), an important precedent has been set, and most local governments will approach more cautiously the use of RF exposure as a regulatory tool for all RF services.

Many jurisdictions are now testing the extent of their ability to regulate *around* the RF exposure issue, too. They might, for instance, require the submittal of very specific showings regarding compliance with the FCC's RFR standard. Or they might specify how and when measurements are to be taken and at what intervals. Even these efforts would appear to conflict with the federal preemption and may not withstand a challenge in the courts. They risk encouraging poor engineering practice by regulating the manner in which engineers might normally conduct the measurements. In total, however, they represent so slight an additional burden that a challenge would likely be made only in a particularly egregious case.

A favorite gambit, one that might not at first seem contrary to federal law, is a moratorium on, often, even the *acceptance* of an application for a new telecommunications facility. An emergency freeze by the appointed Planning Commission for 30 days, an extension for 180 days to study the issue, another extension to formulate proposed policy, another for public comment, another for reformulation before submittal to the elected body with real jurisdiction—the cumulative result of these is an effective block for several years, in some cases, on improvements to existing wireless systems and on deployment of new ones. While no would-be applicant has tested this strategy yet in the courts, private legal opinions have suggested that the legality of such moratoriums would not be sustained.

The natural inclination of virtually all government officials, of course, whether elected, appointed, or career, is to regulate, and there remain plenty of other issues by which local jurisdictions may act on those inclinations. Zoning and land-use restrictions, and other explicitly aesthetic considerations, can be just as effective, and perhaps even keep the issue closer to its essence, which is, "not in my backyard!"[13]

Conclusion

It is hoped that the reader of this text has gained an increased understanding of the issues behind the radiation and regulation of radio frequency electromagnetic fields. The existence and nature of biological effects are the subject of ongoing scientific inquiry. The promulgation of exposure standards is a matter of ongoing bureaucratic debate. The compliance with those standards is a challenge for ongoing engineering development. The regulation of transmitting facilities is a field of ongoing political struggle. Thus, while each of these four elements has seen considerable progress, with few surprises expected now on any front, the field is far from static, and the future may yet hold significant change.

[13]This is the so-called NIMBY syndrome ("Not In My Back Yard"). It is always a chuckle to hear neighbors say, "I couldn't get by without my cellular phone, but don't put a new tower where I can see it!" It is said that there is a range of reasonableness within the groups opposing deployment, from NIMBY through BANANA ("Build Absolutely Nothing Anywhere Near Anything") and ending with NOPE ("Not On Planet Earth").

Appendix

Answers to Commonly Asked Questions

Q: How does the FDA limit for leakage from microwave ovens compare to ANSI82? To ANSI92?

A: The FDA limits for leakage from microwave ovens operating at 2,450 MHz were adopted in 1968: 1 mW/cm^2 at the time of manufacture and 5 mW/cm^2 thereafter, as measured at 5 cm from the unit. This latter value equals the 1982 ANSI Standard. ANSI 1992 specifies 1.63 mW/cm^2 and the 1986 NCRP report specifies 1 mW/cm^2. (pp. 42, 50, 56, 79)

Q: Are there any medically qualified people on the ANSI committee, or was the standard drafted by engineers?

A: There were, indeed, medically qualified persons serving on the ANSI committees. In fact, of the 125 people on the committee at the time ANSI C95.1-1992 was adopted, only about one in eight came from industry. The remainder were affiliated with academic research, civilian government, or the military. (pp. 75–76)

Q: How does the exposure from a typical police, fire, or local government base station compare to that of a typical 900 MHz cellular radio site or a typical 1.9 GHz PCS site?

A: Municipal emergency services generally operate at lower frequencies but at higher powers, since they are intended to cover a larger area than a single cell for telephone or PCS use. Consequently, the exposure levels from municipal systems tend to be higher, although their usage patterns may be more sporadic. Due to typical choices for antenna placement, however, both types of services generally meet the prevailing exposure standards in publicly accessible areas by many thousands of times.

Q: Radiation is just so scary. Isn't letting cellular telephone into our neighborhood like bringing in a little piece of Chernobyl?

A: *Radiation* is a term meaning, simply, that something is spreading out—something is being dispersed (sound waves are an example, as are heat and light from a regular light bulb). *Radioactivity,* as might be associated with nuclear weapons or power plants (such as Chernobyl), refers to something else entirely: the radiation of atomic particles, some of which can cause severe health problems, including death, if absorbed in sufficient quantity. There is no radioactivity associated with cellular telephone base stations or with any other energy source at nonionizing frequencies. (p. 5)

Q: Why does [our local power company] recommend "prudent avoidance" yet they keep adding cellular base stations in the area?

A: The matter of exposure to energy at extremely low frequencies (ELF) does not fall within the expertise of the author, but the frequencies in use by the two different industries are quite different and are acknowledged to interact with humans in different manners. There are presently no safety standards limiting human exposure to the ELF energy associated with power lines. Prudent avoidance in the face of an unknown potential hazard is always sound advice. At radio frequencies (RFs), however, the matter has been researched for decades, and the hazards that can be associated with RF at high power levels are believed to be well understood. Therefore, the safety factors incorporated into the prevailing standards should be adequate and no further avoidance is necessary. (pp. 10–12, 21)

Q: Why do cellular systems need so many base stations?

A: If there were enough frequencies for each cellular telephone to have its own, only a few, centrally located base stations would be required. However, since there are only a few frequencies available, and since there could be tens of thousands of cellular phones in an urban area, the same frequencies must be used over and over again. This is accomplished by establishing small areas, called *cells,* that should have a low enough number of callers on the system so that each can have one of the available frequencies. If too many callers want to use their phones within one cell, they will get busy signals, and the telephone company will get complaints, leading them to squeeze in an additional cell, making all the ones around it just a little smaller. (pp. 167–170)

Q: What is the difference between analog and digital cellular communications?

A: Analog means that the voice is encoded on the radio wave as a continuously varying signal containing tone, pitch, and volume information. In a digital scheme, the same information is broken down into only a few values (0 and 1 in the basic case). The digital data can later be reconstructed back into sound, which humans hear as an analog signal. Some information is lost by the analog-to-digital-to-analog process, but the advantage of digital signals is that they can be packed more tightly, allowing several digital signals (six, often) in the space of one analog signal. (pp. 37–38)

Q: *Do digital cell phones cause health problems that analog ones do not?*

A: Because of their greater efficiency, portable handsets that use digital transmission schemes often turn themselves on to transmit only in short bursts of information. There is no evidence that the energy absorbed by the human body from this action affects us any differently from the continuous energy of an analog unit. Apparently, though, the rate at which digital handsets cycle on and off is similar to the rate at which certain brands of pacemakers send signals internally to their various parts. Current research seems to indicate that holding a digital handset against the chest of someone with a pacemaker can, depending on the brand of the pacemaker, affect the performance of that implanted device, and certain brands of hearing aids are also susceptible to interference.

Q: *Can radio waves from base stations harm me or my family?*

A: There is no evidence to suggest that this is the case. At distances beyond about 25 ft, the fields from most cellular base stations have fallen below the most restrictive public exposure standards. Thus, there is already at least a 50-times safety factor in the radio frequency power levels at that distance. In addition, the total implied safety factor goes up by a factor of 4 every time that distance doubles and by a factor of perhaps 10 for each wall or roof that the signal must travel through. (pp. 6, 21)

Q: *Are there safety standards for exposure to the radio waves from base stations?*

A: Yes, there are two similar standards in active use in the United States. The most recently adopted is ANSI C95.1-1992, which applied a 50-times safety factor to the threshold of biological effects to establish maximum levels for public exposure. NCRP #86 (1986) also incorporates a 50-times safety factor in its maximum public exposure levels. (pp. 56, 79)

Q: *How does exposure to radio waves from base stations compare to that permitted by applicable safety standards?*

A: Exposure conditions near most cellular base stations are hundreds or thousands of times less than the levels allowed by safety standards for continuous exposures. Those standards already incorporate a 50-times safety factor, for total factors of thousands, tens of thousands, or hundreds of thousands of times.

Q: *Can radio waves from a base station affect my________[telephone, answering machine, stereo system, radio tuner, television set, etc.]?*

A: The energy emitted by a cellular base station is controlled within very narrow frequency ranges, and interference to consumer electronic equipment is not reported. Such interference can occur, however, from nearby AM, FM, or TV stations with very strong signals.

Q: Can the radio waves from a base station affect medical equipment or monitors?

A: Apparently, there have been reports of interference to sensitive hospital equipment from hand-held radio transmitters brought inside the hospital. Operation of hand-held units is often not allowed in such locations. No such reports have been noted for the outdoor base station transmitters.

Q: How do local zoning codes treat base stations?

A: That is a legal question and so does not fall within the expertise of the author. It appears that conditional use permits are required in most jurisdictions for base stations sited in residential districts but not for those in commercial districts.

Q: How many other companies can we expect to locate antennas at this site?

A: Not all wireless service providers utilize the same system architecture, nor do they all anticipate equal subscriber growth rates. As a consequence, collocation of base stations is not always practical. In those situations where the providers can collocate, their number will be limited by the total frequencies allotted by the FCC for those services. Only about seven or eight providers in any one market can be accommodated within the present allocation schemes. (p. 170)

Q: Will the construction of a base station reduce residential property values nearby?

A: This question is also beyond the expertise of the author. Most base stations add a negligible amount of RF energy to the neighborhood, and many base stations are installed in a manner so unobtrusive that there is also negligible visual impact. If cell sites did have an adverse impact, properties would be identified by their proximity to them.

Q: Should a base station be located near a school or other potentially sensitive area?

A: Since most base stations meet all applicable safety standards, there is no scientific justification for restricting their location near any sensitive area. Of course, there may be other reasons for doing so, such as minimizing public controversy.

Q: What is a radio wave?

A: Like energy emitted from a light bulb as heat and light, energy emitted by a radio transmitter flows outward spherically in all directions. Radio frequencies are much lower than those of visible light, but many radio services will only perform well when a line of sight exists between the transmitter and the receiver. Data is encoded onto the radio wave in a variety of ways, and appropriate receivers can decode the data as sound or pictures or digital data. (pp. 9, 36–38)

Q: Why do cellular base stations have a limited range?

A: Cellular base stations are designed to have limited range—that is the concept behind a cellular formation and reuse of frequencies. If the base stations had range beyond the neighboring sites, system capacity would be reduced. (pp. 169–170)

Q: Can the radio waves from a base station affect my car's electronic systems?

A: The energy emitted by a cellular base station is controlled within very narrow frequency ranges, and interference to automotive electronic equipment is not reported. In addition, the Society of Automotive Engineers (SAE) has standards limiting the susceptibility of automotive systems.

Q: Can a base station fall over or be blown away in a storm?

A: That is a structural question and so does not fall within the expertise of the author. Most jurisdictions require that construction plans be designed by registered engineers and approved by staff.

Q: How does a base station's visual appearance fit into a community?

A: That is an architectural question and does not fall within the expertise of the author. Most installations on buildings can be designed in an unobtrusive manner.

Q: Isn't it true that airliner pilots develop cataracts as a result of flying near the big TV and radio towers?

A: While this story has been repeated many times, there is no evidence that there is any truth to it. Cataract formation in animals is known to occur at painfully high power density levels, and the exposure conditions for pilots must be millions or billions of times below those. (pp. 21, 24)

Q: Why do antennas have to be so high and so ugly?

A: The antennas need to be relatively high up so that they can "see" beyond the nearby houses and trees into the desired coverage area. The appearance of the antennas can be modified greatly with judicious mounting locations, color selection, and nonmetallic screening.

Q: How close can I really get to the antenna without hurting myself?

A: At the maximum permissible exposure level of the safety standards, there is implied a 50-times safety factor. The standards are based upon "no effects" threshold, not a "no harm" threshold. This means that even in a field 50 times stronger, there is still no evidence of harm. In fact, it may be difficult to experience such a high field, even by approaching and touching the antenna. (p. 21)

Q: What happens to me if I walk through an area marked and signed as exceeding the exposure guidelines?

A: In all likelihood, such an event may not even involve exceeding the guidelines. Many areas marked in this fashion have been identified by predicting the RF power density levels, a practice known to be conservative. Even if the area was identified by measurement, it was probably done when all transmitters were activated, a worst-case condition that may not exist normally. Finally, the common standards both have time-averaging provisions: 6 min for occupational exposures and 30 min for public exposures. Thus, simply walking through a marked area should not be a cause for concern. (p. 126–128)

Q: Why doesn't the telephone company do this instead of some unknown company?

A: The competition within the field of wireless telecommunications is carefully managed by the U.S. government. Due to concerns about excessive concentration of power, certain frequencies are reserved for companies that are specifically *not* the telephone companies.

Q: Why should we put up with this new tower just so some rich guy can use his cell phone?

A: The tremendous growth of wireless telephone services indicates that they are increasingly in demand by people from many walks of life. The services also provide an important emergency response capability that extends the reach of existing municipal services.

Q: The proposed power levels seem low, but how do we know the companies won't go to high power when they have more customers?

A: The proposed transmitters may not be able to produce much more power. Even if they could, there is no incentive for the wireless service provider to increase operating power. The capacity of the cell surrounding the site is determined by the number of frequencies available, not the power used. In fact, increasing power would cause the site to interfere with adjacent cells. As public demands for the service grow, additional sites will need to be constructed and the power levels of existing sites are often actually reduced in order not to cause interference with the new sites. (p. 169)

Q: Isn't a microwave oven the same as a cellular transmitter. How can I know that either one is safe?

A: A microwave oven operates at a frequency of 2,450 MHz, PCS at 1,900 MHz, and cellular at about 900 MHz, so they do use similar frequencies. The biggest distinction is in design: a microwave oven concentrates its power, while the telecommunications services disperse theirs. Conditions inside a microwave oven simply cannot be duplicated at wireless base stations, where exposure levels are generally thousands of times below the safety standards.

Q: I know not to put my kitten in the microwave, even if she were wet. How close to the cellular antennas on the roof should I allow any of my pets to get?

A: The safety standards for humans have been developed largely from testing on animals, and the prevailing standards should be adequate for pet protection, too. Levels had to exceed the public standards by about 50 times before effects were noted in the animals. Appropriately, one of the first effects was a desire to leave the area of the high field; this would appear to provide additional assurance that wireless installations should pose no danger to pets. (pp. 25, 56)

Q: What guarantee do we have that the projected radiation levels will match actual ones at the moment of activation and well into the future?

A: Measurements of the actual power density levels can be made following activation, to confirm the conservative nature of the calculation methodology. Once wireless sites are built, the maximum power levels produced remain essentially constant. Only a change in conditions, such as the addition of another service nearby, would cause the RF fields there to change.

Q: What are the cumulative effects of having several wireless service providers locate their antennas at the same site?

A: *Cumulative* can imply that effects build over time, which is not the case. The better term would be *additive,* which describes the fact that two transmitters, each generating, say, 1 percent of the exposure limit at a particular location, would yield a total exposure there of 2 percent of the limit.

Bibliography

Adair, Robert K., "Effects of Weak High-Frequency Electromagnetic Fields on Biological Systems," *Proceedings of NATO Advanced Research Workshop,* Rome, Italy, May 17–21, 1993.

American National Standards Institute, Standard C95.1-1974, *Safety Level of Electromagnetic Radiation with Respect to Personnel,* Institute of Electrical and Electronics Engineers, New York, 1966.

———, Standard C95.1-1982, *Safety Levels with Respect to Human Exposure to Radio Frequency Electromagnetic Fields, 300 kHz to 100 GHz,* ANSI, New York, 1982.

Clemmensen, Jane M., *Nonionizing Radiation—A Case for Federal Standards?,* San Francisco Press, San Francisco, 1984.

Contemporary Authors, New Revision Series, Gale Research Co., Detroit, 1986.

Dorf, Richard C. (ed.), *The Electrical Engineering Handbook,* CRC Press, New York, 1993.

Egan, Bob, "Wireless Data Communications—A Technology Primer," Digital Equipment Corporation, World Wide Web, 1996.

Elder, J. A., and D. Cahill, "Biological Effects of Radiofrequency Radiation," National Technical Information Service, Report No. EPA-600/8-83-026F, Springfield, Va., 1984.

Environmental Protection Agency, *An Engineering Assessment of the Potential Impact of Federal Radiation Protection Guidance on the AM, FM, and TV Broadcast Services,* Las Vegas, April 1985.

European Committee for Electrotechnical Standardization, Prestandard No. 50166-2, *Human Exposure to High-Frequency Electromagnetic Fields (10 kHz to 300 GHz),* Brussels, Belgium, 1994.

Federal Communications Commission, *Form 303, Application for Renewal of License,* Government Printing Office, Washington, D.C., 1985.

———, Office of Science & Technology, Bulletin No. 65: *Evaluating Compliance with FCC-Specified Guidelines for Human Exposure to Radiofrequency Radiation,* Washington, D.C., 1985.

Federal Communications Commission Public Notice, *Further Guidance for Broadcasters Regarding Radio Frequency Radiation and the Environment,* Washington, D.C., 1986.

Federal Register, "Environmental Protection Agency," vol. 51, p. 27318, Washington, D.C., July 30, 1986.

Grandolfo, Martino, "The Standardization Agreement of the Protection of NATO Personnel against Radiofrequency Radiation," *Proceedings of NATO Advanced Research Workshop,* Rome, Italy, May 17–21, 1993.

Guy, Arthur W., "Non-Ionizing Radiation: Dosimetry and Interaction," *Proceedings of the Non-Ionizing Radiation Symposium,* ACGIH, 1979.

Hammett, W. F., "Meeting the New ANSI C95.1-1992 Requirements," National Association of Broadcasters Engineering Conference, April 21, 1993.

———, *et al.,* "Chapter 2.9, Human Exposure to RF Radiation," National Association of Broadcasters *Engineering Handbook,* 8th ed., NAB, Washington, D.C., 1992.

Hausmann, Erich, and Edgar P. Slack, *Physics,* D. Van Nostrand Co., New York, 1939.

Health and Welfare Ministry, Bureau of Radiation and Medical Devices, Safety Code 6, *Limits of Exposure to Radiofrequency Fields at Frequencies from 10 kHz–300 GHz,* Ottawa, Canada, 1993.

Hitchcock, R. Timothy, and Robert M. Patterson, *Radio-Frequency and ELF Electromagnetic Energies,* Van Nostrand Reinhold, New York, 1995.

Hutchinson, Charles L. (ed.), *The ARRL 1985 Handbook for the Radio Amateur,* ARRL, Newington, Conn., 1984.

Institute of Electrical and Electronics Engineers, Standard C95.1-1991, *Safety Levels with Respect to Human Exposure to Radio Frequency Electromagnetic Fields, 3 kHz to 100 GHz,* IEEE, New York, 1992.

———, Standard 100-1984, *IEEE Standard Dictionary of Electrical and Electronics Terms,* Frank Jay (ed. in chief), 3d ed., New York, 1984.

———, Standard C95.3-1991, *Recommended Practice for the Measurement of Potentially Hazardous Electromagnetic Fields—RF and Microwave,* IEEE, New York, 1992.

Jeppesen, *Federal Aviation Regulations,* Jeppesen Sanderson, Englewood, Colo., 1996.

Johnson, Richard C., and Henry Jasik (eds.), *Antenna Engineering Handbook,* McGraw-Hill, New York, 1984.

Mumford, W. W., "Some Technical Aspects of Microwave Radiation Hazards," *Proceedings of the IRE,* February 1961.

National Council on Radiation Protection and Measurements, *Radiofrequency Electromagnetic Fields, Properties, Quantities and Units, Biophysical Interaction, and Measurements,* report no. 67, Washington, D.C., 1981.

———, *Biological Effects and Exposure Criteria for Radiofrequency Electromagnetic Fields,* report no. 86, Bethesda, Md., 1986.

National Radiological Protection Board, vol. 4, no. 5, *Restrictions on Human Exposure to Static and Time Varying Electromagnetic Fields and Radiation,* Chilton, Didcot, Oxon, England, 1993.

Office of Research and Development, *Evaluation of the Potential Carcinogenicity of Electromagnetic Fields,* report no. EPA/600/6-90/005B), Environmental Protection Agency, Washington, D.C., 1990.

Office of Telecommunications Policy, *Fourth Report* on "Program for Control of Electromagnetic Pollution of the Environment: The Assessment of Biological Hazards of Nonionizing Electromagnetic Radiation," Executive Office of the President, U.S.A., Washington, D.C., June 1976.

Office of the Federal Register, *The United States Government Manual,* U.S. Government Printing Office, Washington, D.C., 1990.

Olsen, R. G., and Barry J. Van Matre, *Measurements of Ankle SAR and Body-to-Ground Current in a Suit-Protected Human Model for Near-Field Exposures, 2–400 MHz,* Naval Aerospace Medical Research Laboratory, Pensacola, Fla., August 1993.

Osepchuk, J. M. (ed.), *Biological Effects of Electromagnetic Radiation,* IEEE Press, New York, 1983.

Polk, C., and E. Postow, *Handbook of Biological Effects of Electromagnetic Fields,* CRC Press, Boca Raton, Fla., 1986.

Prausnitz, S., and C. Sussking, "Effects of Chronic Microwave Radiation in Mice," *IRE Transactions of Biomedical Electronics,* 1962.

Short, J. G., and P. F. Turner, "Physical Hyperthermia and Cancer Therapy," *Proceedings of the IEEE,* January 1980.

Silva, Jeffrey, "Study Says Short-Term RF Exposure Doesn't Increase Mortality," *Radio Communications Report,* April 15, 1996.

———, "True Goals of WTR Questioned by Industry," *Radio Communications Report,* May 6, 1996.

Stein, Jess (ed.), *The Random House Dictionary of the English Language,* Random House, New York, 1967.

Stremler, Ferrel G., *Introduction to Communication Systems,* Addison-Wesley, Reading, Mass., 1990.

Tell, R. A., *Engineering Services for Measurement and Analysis of Radiofrequency (RF) Fields,* FCC RFP no. 94-92, June 1, 1995.

———, and F. Harlen, "A Review of Selected Biological Effects and Dosimetric Data Useful for Development of Radiofrequency Safety Standards for Human Exposure," *Journal of Microwave Power,* December 1979.

United States Bureau of the Census, *Statistical Abstract of the United States,* U.S. Department of Commerce, Bureau of the Census, Washington, D.C., 1995.

United States of America Standards Institute, Standard C95.1-1966, *Safety Level of Electromagnetic Radiation with Respect to Personnel,* USASI, New York, 1966.
Westman, H. P. (ed.), *Reference Data for Radio Engineers,* 4th ed., International Telephone and Telegraph Corporation, New York, 1956.
Wireless Technology Research, L.L.C., *Report on Phase One: Laying the Foundation,* WTR, Washington, D.C., 1995.
World Almanac Books, *The World Almanac and Book of Facts 1996,* New York, 1995.
World Health Organization, Environmental Health Criteria 137, *Electromagnetic Fields (300 Hz to 300 GHz)*, Geneva, 1993.
Yost, Michael G., *Nonionizing Radiation Questions and Answers,* San Francisco Press, San Francisco, 1993.
Zumdahl, Steven S., *Chemistry,* D.C. Heath & Co., Toronto, 1986.

Index

ABOUT THE AUTHOR

William F. Hammett, P.E. is President of Hammett & Edison, Inc. in San Francisco, California, an engineering firm whose specialties include the mitigation of RF exposure conditions. He previously worked in the Corporate Engineering Department of Standard Oil of California, and also at Dean Witter Reynolds in San Francisco. Mr. Hammett has spoken on RF exposure at annual meetings of the National Association of Broadcasters (NAB), has contributed chapters on this topic to the *NAB Engineering Handbook*, Eighth Edition, and *The Electronics Handbook*, and testifies frequently before local governmental agencies.